D0930629

THE PLANET JUPITER

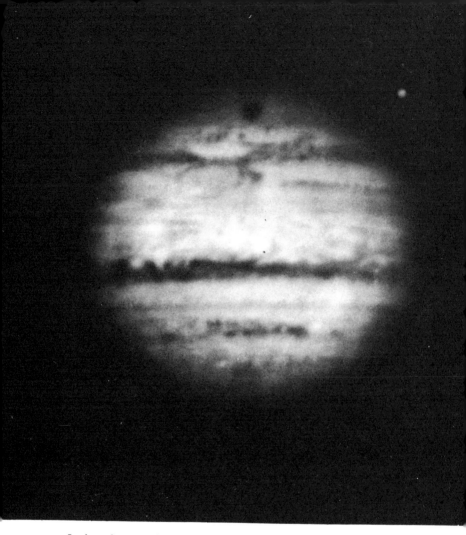

Jupiter photographed in red light. 200-inch Hale Telescope, 1952 Oct. 24.
Photograph from the Mount Wilson and Palomar Observatories.
Satellite III and its shadow well shown

THE
PLANET JUPITER

The Observer's Handbook

by BERTRAND M. PEEK

M.A., F.R.A.S.

with a Foreword by Patrick Moore

FABER AND FABER

LONDON   BOSTON

First published in 1958
This revised edition published in 1981
by Faber and Faber Limited
3 Queen Square London WC1N 3AU
Printed in Great Britain by
Ebenezer Baylis and Son Limited
The Trinity Press, Worcester, and London
All rights reserved

© *Bertrand Meigh Peek 1958*
Revisions © *Patrick Moore 1981*

British Library Cataloguing in Publication Data

Peek, Bertrand M.
  The Planet Jupiter. – 2nd ed.
  1. Jupiter (Planet)
  I. Title
  523.4′5   QB661

  ISBN 0–571–18026–4

CONTENTS

5

CONTENTS

PART III.   THE SATELLITES

ILLUSTRATIONS

DIAGRAMS

FOREWORD

PATRICK MOORE

Bertrand Peek was one of the greatest of all observers of the planet Jupiter in the pre-Space Age. He was an expert observer, and also a first-class organizer; for years he was Director of the Jupiter Section of the British Astronomical Association, and he led a team known all over the world for its energy and skill. During his lifetime, studies of the surface of Jupiter were largely left to amateur astronomers—who acquitted themselves well.

In 1958 Peek produced his book *The Planet Jupiter*. It was at once recognized as filling a most important gap in astronomical literature, and the serious planetary observer regarded it as indispensable. In writing it Peek drew not only upon sources collected in past literature, but upon his own immense experience; and the book was accepted as the standard work not only by amateur astronomers, but also by professionals. It was in a class of its own, and remained the standard even after the end of Peek's Directorship of the Jupiter Section. It was used just as extensively by observers of the planet, and was referred to constantly by Peek's distinguished successors in the Directorship, notably A. F. O'D. Alexander and W. E. Fox.

When the book finally went out of print, around 1962, copies of it were eagerly sought; every Jupiter observer was anxious to have one, and the question of revision was raised time and time again. But Peek had died, and, of course, there were striking developments in planetary research as a whole. The Space Age began on 4 October 1957, with the launch of Russia's artificial satellite Sputnik 1; Yuri Gagarin made his pioneer trip beyond the Earth in 1961; and in 1962 Mariner 2 became the first successful planetary probe, by-passing Venus at less than 22,000 miles and giving us our first reliable information about that decidedly peculiar planet. Mars was by-passed by Mariner 4 in 1965, and clearly it was only a question of time before probes were dispatched towards Jupiter and the other giant planets. Professional astronomers turned back to observations of the planetary surfaces, and Jupiter was studied intensively, mainly by photographic methods.

Yet this did not lessen the value of amateur work. Remember,

Jupiter's surface is always changing, and no professional worker is inclined to glue his eye to the telescope for hour after hour, noting the various features as they drift across the Jovian face and timing their transits across the central m ridian. Peek's book was still needed. Unfortunately, as we all know, inflation made itself all too obvious. For this and other reasons, there were continuing delays in revising the book. Only now is it possible to issue a new edition.

When I was invited to examine the situation with a view to re-publication (on a purely honorary basis, I hasten to add; I regarded it as a great honour), some fundamental decisions had to be made. Peek's original intention, as he had stated in his Preface, was to make the book as complete as possible up to the year 1947, when he virtually retired from systematic observing. This he achieved, though he added various sections to deal with later work—such as the wholly unexpected discovery of radio emissions from the planet. The immediate temptation was to try to continue this policy and bring the book up to date for the 1970s, but there were serious difficulties in the way. Much of the text would have had to be revised, which would entail complete resetting, and the cost would have been more or less prohibitive. Though the book must remain of great importance to the practical observer, it could never become a 'best seller' in the accepted sense of the term, and nowadays the only way to keep the price of a technical work down to an acceptable level is to print a very large edition. Since this was impracticable, there had to be a compromise. Moreover, the observational sections of the original edition remain an excellent guide to the telescope-user. Extending the survey up to the present time would, of course, add to the overall value, but perhaps not to the extent which might be imagined at first glance. The basic principles of Jovian observation remain unchanged.

This does not apply to theoretical studies, discussed by Peek in the third section of the original book. We know a great deal more about Jupiter today than we did in Peek's lifetime, and, inevitably, some of the conclusions which he drew are now known to be completely wrong. A case in point is that of the Great Red Spot, unquestionably the most famous of all features on the planet. Space-probe results have shown that it is in the nature of a whirling storm—a phenomenon of Jovian meteorology. Peek's own theory was quite different, and I quote here from his original Chapter 29:

It was from a suggestion made by Wildt in 1939 that the author developed his tentative explanation of the phenomena exhibited by the Red Spot. In a paper read in February of that year before the American Philosophical Society (*Proceedings*, 81, 135) Wildt put forward the hypothesis that the Spot is due to the presence of a large solid body, whose length and breadth are comparable with those of the Spot as we see it and whose depth may be expected to be of the same order of magnitude as its breadth. This object he imagined to be floating, not in a liquid but in an ocean of highly compressed permanent gases. It is not stated in the paper whether what is seen from the Earth is considered to be the actual upper surface of the solid; but that would correspond to the theory of the present author, which is set out in detail below.

Consider first the well-known experiment of immersing an egg in a solution of salt and water. If the solution is more concentrated toward the bottom of the containing vessel, as it is likely to be at first, the egg, while remaining completely under water, will float at a level determined by its density. Now replace the solution by Jupiter's atmosphere, of which the density increases rapidly with the depth until it almost certainly approaches that of the liquid state, and let the egg be represented by some solid whose upper surface lies at least some tens of kilometres below the top of the cloud layer and whose nature will be considered in a moment. Any influence tending to disturb the equilibrium of the density distribution in the atmospheric layers will bring about a change in the level at which the solid will float and, except at the poles, will cause it to approach or recede from the planet's axis of rotation. . . . It should be easy enough to accept the next postulate, namely, that during the past 120 years its level has been subject to slight variations having a total range not greatly exceeding 10 km. Indeed, if there were no mechanical resistance to changes of motion, the range of 10 km. in depth would suffice to account for all the changes in rotation period and hence for the whole of the drift in longitude of the Red Spot since 1831.

The theory was completely reasonable at the time when Peek proposed it, and it would still be reasonable today but for the results obtained from the Pioneer and Voyager space-probes. Pioneer 10 by-passed Jupiter in December 1973, sending back spectacular pictures as well as an immense amount of information about the planet's composition, magnetosphere and radiation zones; it then began a never-ending journey which will take it out of the Solar System (carrying a plaque which, it is hoped, will enable its planet of origin to be identified by any far-away civilisation which may happen to recover it in the remote future—admittedly a very slim possibility!). Pioneer 11, essentially a twin probe, made its pass in December 1974, confirming the earlier results. Pioneer 11 was then diverted on to a rendezvous with Saturn, and in 1979 sent back the first close-range pictures of that remarkable and beautiful world. It, too, has now embarked upon a journey which will take it permanently out of the Solar System. Next came Voyager 1, which made its rendezvous with Jupiter in March 1979, and Voyager 2, whose closest approach took place in the following July. Both these moved on to rendezvous with Saturn, reaching their target in November 1980 and August 1981 respectively. If all goes well, Voyager 2 will also by-pass Uranus (1986) and Neptune (1989).

The Red Spot, which had been extremely prominent for most of the 1960s and 1970s, was shown in great detail by both Pioneers and Voyagers, and the resolution of the pictures was vastly better than anything obtainable from Earth. Much information was also derived concerning the planet's constitution. According to a model due to J. D. Anderson and W. B. Hubbard in the United States, Jupiter has a relatively small rocky core made up of iron and silicates; the temperature is of the order of 30,000 degrees Centigrade. Around this is a shell of liquid metallic hydrogen, which is in turn overlaid by liquid molecular hydrogen; outside comes the gaseous atmosphere, about 1000 km. deep, made up of 82 per cent hydrogen, 17 per cent helium and a mere 1 per cent of other elements. It contains water droplets, ice crystals, ammonia crystals and ammonium hydrosulphide crystals. Gases warmed by the internal heat of the planet rise into the upper atmosphere and cool, forming clouds of ammonia crystals floating in gaseous hydrogen; these clouds make up the bright zones, which are colder and higher than the dark belts. Jupiter radiates more energy than it would do if it depended entirely upon radiation received from the Sun, and it was once thought that this excess energy might be due to a slow contraction of the globe; but as Jupiter is now believed to be mainly liquid, and liquids are incompressible, it is more likely that the cause is heat remaining from the far-off period when Jupiter was formed.

Radio emissions from Jupiter were first detected (more or less accidentally, it must be admitted!) by B. F. Burke and F. L. Franklin in 1955. We now know that there is an extremely powerful and complex magnetic field, and that Jupiter is associated with radiation zones so strong that they would be instantly lethal to any astronauts unwise enough to venture inside them. Indeed, the instruments on Pioneer 10 were almost put out of action by them, and modifications were made to the fly-by paths of the later probes. The Jovian magnetosphere is so extensive that the 'tail' extends out to the orbit of Saturn, and when suitably placed Saturn may lie inside it.

Quite apart from studies of Jupiter itself, the Voyagers produced a major surprise by revealing that the planet has a ring. The ring system of Uranus had been discovered in 1977, when a 9th-magnitude star 'winked' both before and after occultation by the planet, but nothing comparable had been seen with Jupiter before the flights of the Voyagers. The Jovian ring is probably unobservable from Earth; it lies about 57,000 km. above the cloud tops, with a width of 6500 km. and a thickness of a few kilometres at most. It cannot be compared with the rings of Saturn, and presents theorists with a number of problems which remain to be solved.

Then, of course, there are the satellites. At the time of writing fifteen are known, the last two having been discovered by space-probes, but only four are large. These are Io, Europa, Ganymede and Callisto, known commonly as the Galileans because they were studied by Galileo in the very earliest days of telescopic research. For many years the names were regarded as 'unofficial', and Peek, in his book, generally alludes to them by their numbers, I, II, III and IV. Their physical details are much better known today than they were in Peek's time, and for this reason I have revised the first three pages of his section 'A General Survey of the Satellite System'—the only alterations I have made to his original text. The remaining satellites are extremely small, and may be asteroidal in nature. No doubt further minor satellites await discovery.

The Galileans have proved to be amazingly interesting, and quite unlike each other. Io, the innermost, has provided the greatest surprise of all. The surface is intensely red; there are no craters, but there are active volcanoes, shown by both the Voyagers in states of violent eruption. The material sent out from Io plays an important rôle in the Jovian radiation zones, and goes a long way towards explaining the already-established correlation between Io's orbital position and the radio emissions received from Jupiter. According to Bradford Smith, there is a molten silicate interior overlaid by a silicate crust, which is itself overlaid by a 'sea' of molten sulphur and sulphur dioxide about 4 km. deep, with only the outermost kilo-metre solidified. Since this silicate magma is denser than the liquid sulphur, it cannot reach the surface; liquid sulphur moves outward and reacts with deposits of sulphur dioxide, so that the expanding

13

gas literally explodes into space. Alternatively, Stanton Peale suggests that tidal forces have melted Io's interior, leaving a 20-km. crust which rises and falls tidally. The sulphur dioxide 'atmosphere' has a density of about 10^{-7} that of the Earth's air at sea level, so that by everyday standards it is negligible.

At all events, Io is the only world, apart from the Earth, where active vulcanism has been established. Europa, slightly smaller and less dense than Io, is completely different. There are few or no craters, and the surface is incredibly smooth; it has been claimed that it is 'smoother than a billiard ball', and the surface relief cannot be much more than 50 metres anywhere. The crust is icy, and there are innumerable shallow cracks. It has been suggested that water from the interior cooled, forming an ice-mantle up to 100 km. thick; the icy surface was fractured, and the cracks filled with dark material from below. Europa may be less spectacular than Io, but in its way it is perhaps even more puzzling.

The two outer Galileans, Ganymede and Callisto, are much larger (comparable with the planet Mercury), and are less dense, with icy surfaces. On Ganymede there is one large, dark, circular feature about 3200 km. in diameter, and there are many craters, together with light, linear stripes. Callisto is the most heavily cratered body known; the features are so crowded that there is hardly room for any more of them, and the whole impression is of exceptionally great age. The contrast between the ancient, dead Callisto and the violently active Io could hardly be more pronounced.

The fifth satellite, Amalthea (discovered by E. E. Barnard in 1892), was also recorded by the Voyagers. Like Io, it moves in the midst of the Jovian radiation belts; it is a small, red, irregularly-shaped body. Because of its insignificant size, and its closeness to Jupiter, it is beyond the range of most amateur-owned telescopes. The same applies to the outer asteroidal satellites, four of which (Ananke, Carme, Pasiphaë and Sinope) move round Jupiter in a retrograde direction.

Obviously, the minor satellites appear only as starlike points as seen from Earth, but the Galileans show definite disks, and a certain amount of surface detail had been recorded before the Pioneer and Voyager missions. Maps had been published, notably by French observers working at the high-altitude Pic du Midi station in the Pyrenees. These maps are now wholly obsolete, but I have not altered Peek's references to the Earth-based observations, which are at least of considerable historical interest.

Further probes to Jupiter are being planned, and some will

undoubtedly be launched during the next few years. This being so, it may be asked: What is the point of continuing to study the planet through modest telescopes based on Earth? In fact, observations of Jupiter are as useful now as they have ever been. Remember, the surface is always changing, and one never knows what to expect, as is abundantly clear from a perusal of what Peek has written in his text. If anything dramatic happens, it is the amateur observer who is most likely to detect it at an early stage. Photographic coverage from professional workers can never be complete, and without the visual observer there is always a real chance that something important will be missed.

Moreover, Jupiter is a convenient planet inasmuch as it comes to opposition every year. It may be useful to give the opposition dates from 1980 until the end of the century, with the opposition magnitude:

Year	Date	Diameter, seconds of arc	Magnitude
1980	Feb. 24	44·7	−2·1
1981	Mar. 26	44·2	−2·0
1982	Apr. 25	44·4	−2·0
1983	May 27	45·5	−2·1
1984	June 29	46·8	−2·2
1985	Aug. 4	48·5	−2·3
1986	Sept. 10	49·6	−2·4
1987	Oct. 18	49·8	−2·5
1988	Nov. 23	48·7	−2·4
1989	Dec. 27	47·2	−2·3
1991	Jan. 28	45·7	−2·1
1992	Feb. 28	44·6	−2·0
1993	Mar. 30	44·2	−2·0
1994	Apr. 30	44·5	−2·0
1995	June 1	43·6	−2·1
1996	July 4	47·0	−2·2
1997	Aug. 9	48·6	−2·4
1998	Sept. 16	49·7	−2·5
1999	Oct. 23	49·8	−2·5
2000	Nov. 28	48·5	−2·4

To sum up: in undertaking the preparation of this new edition of Peek's classic book, I have retained all his observational chapters, entirely unaltered, and without extending them to include pheno- mena seen on the planet between Peek's time and the present day, since this would have greatly increased the published price of the book and would not have added so much to its intrinsic value as might be thought. I have deleted those theoretical sections which are now completely out of date, since this would have meant complete rewriting and would hardly be relevant; this is, after all, a book for the practical observer, and all the theoretical information is available in technical periodicals. I have retained the original Preface, apart from deleting three lines which refer to sections now omitted. The only alteration I have made to Peek's original text has been in the introduction to his general survey of the satellite system, purely because only twelve satellites were known when the first edition was completed, and data concerning both the Galileans and the minor satellites have been revised.

Observers of Jupiter will, I know, be glad that the book has become available again. The Giant Planet has lost none of its fascination, and Bertrand Peek will always be remembered as one of its greatest students.

PATRICK MOORE

SELSEY

PREFACE TO THE FIRST EDITION

This book has been written in the hope that it will fill a long-standing gap on the shelves of the astronomer's library. Hitherto no attempt has been made to present in a single volume an account of either the numerous observations that have been made of the planet Jupiter, the facts that have been revealed by these observations or the theories that have been based upon them. The serious student has been compelled, therefore, to browse among the original publications; and authors of text-books on general astronomy, who have not been able to spare the time for such detailed research, have too often resorted to the iteration of a few traditional beliefs about the planet's economy, many of them out of date by half a century, and have then passed on to Saturn, possibly without realising how much they have left unsaid about the greatest planet in the solar system.

It was the author's original intention to make the book as complete as possible up to the year 1947, when he himself retired from making the systematic observations which had occupied so much of his time at the telescope during the previous twenty-five years. But various circumstances combined to delay both the start and at first the progress of the work during a time when activity on Jupiter was by no means at a standstill, with the result that, while completeness to 1947 remained the primary objective, it seemed desirable to include in some of the chapters records of events that had been occurring during the actual writing of the text. The effect of this policy has been to render the scope of the various chapters somewhat uneven with regard to date; but the author feels that in exceeding his original intention he has added considerably to the value of the book, parts of which have even required some modification of the original typescript before they could be sent to the printers.

In recounting the various events that have been recorded on Jupiter's surface the author's mind has often travelled back over the years to the early days of his own observing career, when he was privileged to be associated with a little group of observers, all of them members of the Jupiter Section of the British Astronomical Association, with whom he was continually in contact either personally or by correspondence and from whom he derived much of the enthusiasm which is necessary for the carrying out of an exacting observational programme. The names that are uppermost in his mind are those of Mr. A. Stanley Williams, whom he actually never met, Rev. Theodore E. R. Phillips, M.A., D.Sc., Instructor-Captain Maurice A. Ainslie, R.N., and Mr. F. J. Hargreaves. Of these his greatest debt was to Phillips, who was Director of the Jupiter Section from 1901 until the author succeeded him in 1934; for to work under Phillips' encouragement and guidance was in itself an inspiration. It was he who should have written the first book on Jupiter. But he could never be persuaded to spare the time from his other arduous commitments; and so the task has now devolved upon one of his erstwhile pupils, who can only hope that the master would have commended the results of his labours. These have been largely sustained by the encouragement and co-operation he has received during the last two or three years from Hargreaves, who alone with the author remains of that original little fraternity but who unfortunately no longer has the time to take an active part in the observation of the planet.

The magnitude of the author's debt to Hargreaves is difficult to express. If the value of his co-operation as an observer in the past could be equalled, surely it is only by the interest he has taken in the preparation of this volume. He has not only read, but studied, every word of the manuscript as from time to time the various chapters have been finished, pointing out during the process sundry slips and at least one major error; nor has he hesitated to recommend or to criticise, so that his approval of the whole in its final form does much to sustain the author's hope that something worth while has been accomplished. On top of all this he has insisted on finding the time, of which he has extremely little to spare, for casting an expert eye over the proofs in order to eliminate as far as possible the final traces of error.

The book would have been tne richer, had it been possible to include among the illustrations a greater number of Phillips' beautiful drawings. But it is no longer practicable to reproduce these from the originals; those which do appear have been copied from published reproductions with great care, and he hopes with fidelity, by the author; but it seems inevitable that they must have suffered to some small extent during the double process.

The author wishes to express his appreciation to the Council of the British Astronomical Association for permission to make unrestricted use of the material contained in the publications of the Association. His thanks are also due to the following:

The Director of the Mount Wilson and Palomar Observatories for several of the photographs reproduced in the Plates.

Dr. A. Dollfus, the Director, and M. H. Camichel of the Pic-du-Midi Observatory for the photographs reproduced as Plate XV.

H.M. Stationery Office for permission to reproduce from the Nautical Almanac the Tables given in Appendix III.

The Director, Dr. A. F. o'D. Alexander, and other members of the B.A.A. Jupiter Section, whose names appear at various places in the text, for having kept the author informed of the progress of recent developments in advance of the publication of the observations.

It is to be hoped that the ensuing pages will not only entertain and enlighten the reader, but that in some cases they may kindle in him a genuine desire to see for himself what it is all about and perhaps even to add his own contribution to our present knowledge of Jupiter. It is essential for anyone who feels such an urge to make contact with others who share his interests; and almost the only way to do this is to become a member of an organisation such as the British Astronomical Association, the work of whose members has been so frequently referred to in the text. On no account should diffidence, either with regard to his own ability or because of the lack of instrumental equipment, deter anyone who is interested in astronomy from seeking admission to such a fellowship, where he will quickly receive instruction if he is in need of it and will learn that not only necessity, but even a strong enough desire, may be the mother of a telescope, if there is the likelihood that it will be employed in genuine scientific research.

At the present time there is any amount of scope for the amateur who is prepared to observe Jupiter systematically and the need for co-operation from the southern hemisphere is as great as it has ever been; if, therefore, the perusal of these pages should result in the addition of even one more devotee to the ranks of the really ardent students of the planet, then at least they will not have been written entirely in vain.

HERNE BAY
March 1957

PART I
INTRODUCTION

Chapter 1

GENERAL REMARKS AND PARTICULARS

Jupiter is the largest of the nine great planets that revolve around the Sun, from which it comes fifth in order of distance. Its mass is nearly two and a half times that of all the other planets added together, its nearest rival, Saturn, though inferior by less than twenty per cent in linear dimensions, having not quite three-tenths of the mass of Jupiter. It has often, therefore, been not inappropriately referred to as the Giant Planet.

The distance of Jupiter from the Sun is rather more than five times that of the Earth and it completes one revolution in just under twelve years. Thus it will be seen that the Earth must pass almost directly between the Sun and Jupiter once in about every thirteen months, since the angle between the planes of their respective orbits is only a little more than one degree. At such times the planet will be most favourably placed for observation; for not only will its distance from the Earth be near a minimum but it will be on the meridian at midnight and above the horizon all night long.

Jupiter can be profitably observed for about ten of the thirteen months that elapse between its successive conjunctions with the Sun. These ten-monthly spells are known as apparitions and at the Earth's equator all apparitions are almost equally favourable; in high latitudes, however, the best observing conditions are provided when opposition occurs at mid-winter. In the British Isles, for example, at the time of a December opposition the planet will be above the horizon for about sixteen hours each night, during most of which it will be situated at a considerable altitude, the last consideration being essential if first-class observations are to be made. If, on the other hand, opposition occurs in June, the planet will never attain an altitude of more than about sixteen degrees and the time available for observation, which cannot under the circumstances be of the highest quality, will be confined to an hour or two on either side of meridian passage.

Owing to the rapid rotation of Jupiter upon its axis its polar diameter is less by about one part in fifteen than that measured through the equator. The apparent diameter as seen from the Earth is greatest, of course, when the distance between the Earth and the planet is least and in the equatorial direction may exceed fifty seconds of arc; the mean opposition value of the apparent equatorial diameter, however, is 46·86 seconds and since, moreover, the varying distance of Jupiter is never so great as to give rise to an apparent diameter of less than about two-thirds of this, the disk never appears so small, as unfortunately happens in the case of Mars, that useful observations cannot be made with instruments of moderate size.

The inclination of Jupiter's equator to the plane of its orbit around the Sun is little more than three degrees. In consequence seasonal effects, if any, are exceedingly small, none having been as yet detected. Indeed, on account of its far greater distance, the light, heat and other radiations that fall upon Jupiter from the Sun have on arrival less than one-twenty-seventh of the intensity of those that reach the Earth.

The surface of the planet is surrounded by an atmosphere that is quite impenetrable either visually or photographically; and since the general arrangement of the cloud formations, which are all that we have so far succeeded in actually observing, is at first glance little more than a series of dark and light bands, the uninitiated may well wonder how it is possible to devote the whole of a book of some 280 pages to the discussion of such unpromising material. It is to be hoped, however, that the intending reader will not be deterred from exploring a little further and that not only will his interest and curiosity be aroused and at any rate partially satisfied, but that he may himself begin to appreciate how the remarkable series of phenomena exhibited by Jupiter, ever changing and many of them still quite baffling, has aroused the enthusiasm of observers and theorists alike and has inspired some of them to unremitting efforts in their determination to add to our knowledge of the meteorology and the general constitution of the planet.

We conclude this short opening chapter by tabulating the principal dimensions and other numerical quantities associated with the planet and its orbit. The two remaining chapters of Part I describe the nomenclature employed by students of Jupiter and provide a short glossary of technical terms, while the rest of the book is devoted to presenting as exhaustive a description as possible of the methods and results of observation and of the progress that has been made theoretically in explaining what has been observed.

Numerical Quantities associated with Jupiter and its Orbit

	Equatorial	Polar
DIAMETER		
Kilometres	142,700	133,200
Miles	88,700	82,800
Earth=1	11·2	10·4

MASS	
Grammes	$1·9 \times 10^{30}$
Earth=1	318·4
Sun=1	1/1047·35
DENSITY	
Gm./c.c.	1·34
Earth=1	0·24

	At Equator	At Poles
SURFACE GRAVITY	2·64	2·67
Earth=1		

ROTATION PERIOD	
System I	$9^h\ 50^m\ 30^s\!.003$
System II	$9^h\ 55^m\ 40^s\!.632$
INCLINATION OF EQUATOR TO PLANE OF ORBIT	$3°\!.07$
MEAN DISTANCE FROM SUN	
Kilometres	$7·78 \times 10^8$
Miles	$4·83 \times 10^8$
Earth=1	5·2028
SIDEREAL PERIOD OF REVOLUTION IN TROPICAL YEARS	11·8622
MEAN SYNODIC PERIOD (opposition to opposition) IN DAYS	398·88
ECCENTRICITY OF ORBIT	0·04843
INCLINATION OF ORBIT TO ECLIPTIC	1° 18′ 20″

CHAPTER 2

NOMENCLATURE

Figure 1 on this page shows diagrammatically the aspect of Jupiter's surface. The dark strips are called Belts and the light ones Zones, except that the thin grey line sometimes seen threading the middle of the Equatorial Zone is known as the Equatorial Band. In the illustration north is at the bottom, as it is seen with an inverting telescope by an observer in the northern hemisphere.

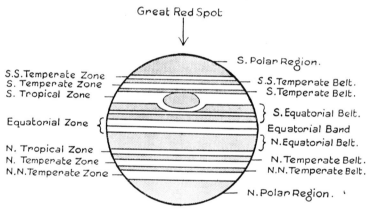

FIGURE 1.—Diagram illustrating the nomenclature adopted in this book.

The four main belts nearest to the equator, though subject to changes of width and intensity, are permanent features of the planet. Their names, beginning with the most northerly, are: the North Temperate Belt (N.T.B.), the North Tropical Zone (N.Trop.Z.), the North Equatorial Belt (N.E.B.), the Equatorial Zone (E.Z.), the South Equatorial Belt (S.E.B.), the South Tropical Zone (S.Trop.Z.), and the South Temperate Belt (S.T.B.). The abbreviations in brackets are standard and will be used extensively in this book. A suffix, n or s, after the abbreviation denotes the north or south edge of the belt referred to, e.g. N.E.Bn., N.T.Bs.; but in the case of a zone, e.g. E.Zn., the north or south *half* of the zone is usually implied. It will be seen that two separate components of the S.E.B. are shown; this

26

represents its most usual appearance. The components are generally designated N.comp.S.E.B. and S.comp.S.E.B. or, when the edges are clearly not intended, S.E.Bn. and S.E.Bs.; e.g. a note reading 'S.E.Bs. invisible' can only mean that the whole of the S. component of the belt was missing. Other belts, notably the N.E.B. and S.T.B., may sometimes show more than one component, when a similar notation may be employed.

Northwards from the N.T.B. we have the N. Temperate Zone (N.T.Z., sometimes N.Temp.Z.), the N.N. Temperate Belt (N.N.T.B.), the N.N. Temperate Zone (N.N.T.Z.), the N.N.N. Temperate Belt (N.N.N.T.B.) and so on till we reach the North Polar Region (N.P.R.), which is the dusky hood surrounding the north pole of the planet. In the southern hemisphere the nomenclature is exactly similar. The number of belts visible between either of the Temperate Belts and the corresponding Polar Region is very variable and at times even the N.N.T.B. or the S.S.T.B. may become merged in a general polar duskiness, the edge of which is separated from the Temperate Belt by only a single Zone, the N. or S. Temperate; yet as many as five 'Ns' or 'Ss' have occasionally been found in the records, as a prefix to a 'T.B.'.

Now and then two belts on the poleward side of one of the Temperate Belts may lie so close together that it is difficult to know whether to name them separately or to classify them as two components of a single belt. Unfortunately latitude measurements of these high-latitude belts are difficult to obtain at the telescope, since their faintness may allow them to be obliterated by the web of a filar micrometer. As we are dependent therefore upon drawings, which are eye estimates only, for a record of their positions, their exact latitudes and therefore their identification from apparition to apparition must be subject to some uncertainty.

The dusky oval object shown in the diagram as occupying the S.Trop.Z. and indenting the S. edge of S.E.B. is the famous Great Red Spot (R.S.) and the 'bay' it forms in the S.E.Bs. is known as the Red Spot Hollow (R.S.H.). Since systematic observations were first begun, if the Red Spot itself has faded to invisibility, as it often does, the Hollow has always remained to indicate its longitude; only very rarely have both been invisible together.

From 1901 until about 1940 there was another independent portion of the S.Trop.Z. that showed a duskiness of varying extent and intensity, so that there appeared a brighter section and a darker section of the zone according to the longitude presented. The darker region has been named the South Tropical Disturbance (S.T.D.).

CHAPTER 3

GLOSSARY OF ASTRONOMICAL TERMS

To give detailed definitions of all the scientific terms used in this book would be too heavy an undertaking; as, however, the vocabulary of the astronomer contains many words and phrases that are used far more rarely, if at all, by workers in other branches of science, a short glossary is given below to explain the meanings of certain expressions that may be found in the text of this book and might otherwise be puzzling even to the trained scientist who has only a slender acquaintance as yet with astronomical literature. Since the main object is to enlighten the uninitiated, strict scientific accuracy has not in every case been observed; to give, for example, a definition of 'opposition' that would be commended by the Superintendent of H.M. Nautical Office would be to introduce further technical terms, whereas for most purposes all that the reader needs to know is that three objects are more or less in a straight line. That being understood, we proceed with the definitions.

Albedo. When an object is illuminated by an external source, it is seen by the light which is scattered from its surface. The ratio of the total light scattered by a given area to the total light falling upon it is known as the albedo of the surface. The albedo is thus a measure of the relative brightness of two or more surfaces that are situated at equal distances from the same source of illumination.

Aperture. The aperture of an astronomical telescope is the linear diameter of the objective lens or mirror that forms the primary focal image.

Apparition. The interval between successive conjunctions of a planet with the Sun during which the planet is favourably placed for observation.

Barlow Lens. A compound achromatic negative lens that may be placed between the objective and the primary focus of a telescope. Its effect is to decrease the angle of the converging cone of rays and thus to displace the focal plane to a somewhat greater distance from the objective. If F inches is the focal length without the Barlow lens and if the effect of placing the Barlow x inches inside the original

28

focus is to transfer the final focus to a distance of y inches beyond the Barlow, the equivalent focal length of the system is Fy/x inches. This factor also represents the linear increase in the size of the primary image, whence we see that by placing a suitably designed Barlow quite close to the original focal plane considerable increase in magnification may be obtained with great economy of tube extension.

Conjunction. The configuration of Earth, Sun and Planet, when the three lie most nearly in a straight line with the Earth at one extremity. If the planet is then between the Earth and the Sun, the planet is said to be in inferior conjunction; if beyond the Sun, in superior conjunction. Only superior conjunctions of Jupiter can occur, since it is always farther than the Earth is from the Sun. Any two celestial objects are said to be in conjunction when, as seen from the Earth, their relative motion has reduced their apparent distance from each other to a minimum; more technically, conjunction is reached when their longitudes, measured on the ecliptic, are identical. The term may also be applied to two surface markings that have arrived at the same longitude on Jupiter.

Declination. The equivalent on the celestial sphere of latitude on the surface of the Earth; it is, of course, an angle and is measured north and south from the celestial equator. If the declination of a heavenly body and the latitude of a terrestrial station are the same, the former will pass daily directly overhead when viewed from the latter. Compare 'Right Ascension' below.

Definition. The quality of the telescopic image which is dependent on the optical steadiness of the Earth's atmosphere, through which the light from a celestial body must pass before it enters the instrument of an observer. The term may be used to refer to the state of the air, whether or not observations are actually in progress. Really first-class definition is experienced all too rarely in temperate climates. Some of the steadiest air is to be found over plateaux at considerable altitudes above sea level under desert or semi-desert conditions. It is not uncommon, however, for distortion of the image to be due to currents of air originating in the observatory dome or even in the apparatus itself.

Dispersion. The resolution of composite light, e.g. white light, into discrete wavelengths (colours), either deliberately, by means of a prism or grating, or unavoidably, as when the light passes through a lens or even through the Earth's atmosphere.

Eclipse. The total or partial disappearance or fading of a heavenly body owing to the source of its illumination being cut off by another body into whose shadow it passes. In this sense an eclipse of the

Sun is not strictly an eclipse but an occultation—see below; an eclipse of the Moon, however, is a true eclipse.

Ecliptic. The plane of the Earth's orbit round the Sun. Also the apparent path among the stars followed accurately by the Sun and approximately by the planets.

Ephemeris. A table of predictions of astronomical phenomena or of the future positions and configurations of planets, comets and other moving celestial objects. Plural—ephemerides.

Following. See 'preceding', below.

Limb. The true edge or profile of Sun, Moon, planet or satellite as presented to the observer—compare 'terminator', below.

Occultation. The obscuring of one heavenly body by the actual passage of a second body between the first and the observer.

Opposition. The configuration of Earth, Sun and Planet, when the three lie most nearly in a straight line with the Earth in the middle. When a planet is in opposition, the Earth, as seen from the planet, would be in inferior conjunction with the Sun. Only planets whose orbits are exterior to that of the Earth can therefore come to opposition and be viewed on the meridian at midnight.

Preceding. The words 'preceding' and 'following', abbreviated to p. and f., are used when referring to the directions to and from which the Earth's diurnal rotation causes a heavenly body to appear to move either in the sky or through the field of view of a telescope. They may also describe the sides or edges of a body that appear to lead or follow during such motion; and since the sense of Jupiter's rotation is such that the surface markings are thereby carried in the same direction as its motion through the field, they may be used to describe the relative positions of markings on the disk with less ambiguity than if the words 'west' and 'east' were employed.

Quadrature. The configuration of Earth, Sun and Planet, when the angle Sun–Earth–Planet is most nearly a right angle. Only planets external to the Earth can arrive at quadrature.

Revolution. The progress of a planet, satellite or other heavenly body in its orbit about another body, e.g. of the Moon about the Earth. The word should not be used to describe rotation about an axis.

Right Ascension. The equivalent on the celestial sphere of longitude on the surface of the Earth. The celestial meridian from which it is measured towards the east, usually in time but sometimes in angle, passes through the celestial poles and through the point among the stars where the Sun crosses the equator at the vernal equinox. The complementary coordinate is 'declination', described above.

Rotation. The spinning motion of a body about an axis, not to be confused with 'revolution' as described above. The rotation period of Jupiter is a little under ten hours, its period of revolution nearly twelve years.

Seeing. This expression is synonymous with 'definition' as already described. The layman frequently uses it when referring to the clearness or transparency of the sky; but the astronomer never does this, for 'seeing' may be excellent on a night when haze is so thick that only the brightest stars can be seen.

Terminator. The limit of the visible part of the disk of a planet or satellite, set by the angle at which it is being illuminated by the Sun. When the Moon is at the first quarter, for example, the visible part of the limb is a semicircle, the terminator a straight line. In crescent or gibbous phases the terminator is semi-elliptical, except for irregularities due to lunar surface features.

PART II

OBSERVATIONS OF JUPITER'S SURFACE

CHAPTER 4

THE AIMS, METHODS AND LIMITATIONS OF VISUAL OBSERVATION

It is probable that most amateur astronomers who possess a small telescope of 3 or 4 inches aperture, and who have already become familiar with the exquisite but somewhat static beauty of Saturn, would agree that Jupiter of all the planets is the one that best repays observation. A very small aperture with a magnifying power of 20 or so will show an obvious disk with one or two dusky belts crossing it, while even a good pair of binoculars will reveal the four bright satellites; indeed, claims that the brightest of these, No. 3, has been seen with the naked eye seem to have been established.

With a 2-inch telescope, bearing a magnification of about 60 diameters, the shadows of the satellites may readily be seen as little black dots when they are projected in transit upon the disk of the planet; and a 3-inch, working under good conditions, will reveal that the belts, which in general run parallel to the planet's equator, are not entirely regular but show darker patches or condensations and sometimes projections at their edges. An occasional bright spot may also be detected with a 3-inch, especially if it partly overlaps one of the dark belts. Moreover the constantly changing configurations of the satellites can be followed with quite a small instrument; also their disappearances and reappearances as they enter or leave the shadow of Jupiter when they become eclipsed. It is fascinating to watch a little moon fade from its normal brightness to invisibility in two or three minutes for no apparent reason or, conversely, to see one appear where a moment ago there was no trace of it and rapidly brighten at a considerable distance from the planet's limb or terminator.

35

Apart from the photographic confirmation of contrasts in the colours of some of the surface features, the whole of our present knowledge of the behaviour of Jupiter's spots and markings has been derived from visual observations; and these have been made almost exclusively by amateur astronomers. None of the great observatories has yet undertaken to make a systematic photographic record of the changes that are continually taking place on the planet's surface. An observer and an instrument of at least 30 inches aperture would have to be detailed specially for the task and for months on end would be available for nothing else; moreover, until there is a vast improvement in the speed of fine-grained photographic plates, it would be only on the finest nights that a large telescope could photograph as much detail as could be recorded by a visual observer using a much smaller instrument.

The aims of visual observation of Jupiter may be summarised as follows:

(1) To place on record changes of form, intensity, colour and position of the surface features.

(2) To make the record as continuous and complete as possible, in order that it may furnish reliable information regarding any such changes as may be systematic or periodic.

(3) To provide data which may inspire and against which may be checked physical theories that will ultimately give us a better understanding of the laws that govern the distribution and circulation of the gases that comprise the planet's atmosphere.

Before discussing in detail the methods of observing form and intensity, colour and position, we must consider a very important item.

The Choice of Telescope

In no department does the maxim, 'a good big 'un is better than a good little 'un', carry more weight than in most branches of practical astronomy, among them the visual observation of the details of a planetary surface. A large telescope can be stopped down, should atmospheric conditions demand it, though the benefit to be derived from this procedure is open to question; but it is impossible, however steady the atmosphere, to stretch a 6-inch objective into a 12-inch. Therefore the bigger the better within practical limits, the limits being set by considerations of cost and convenience.

Although something worth recording may be seen even with a 3-inch, the intending student of Jupiter should have available a telescope of not less than 6 inches aperture. With such an instrument a very great deal of first-class systematic work can be accomplished and only the smallest of the really important markings will be beyond its reach; indeed, until only a year or two before his death, the famous Stanley Williams made all his invaluable observations with a 6-inch reflector. An 8-inch is probably adequate for all purposes; a 12-inch certainly is. The bulk of the author's work has been done with a 12-inch reflector; and, although it would not be true to say that he has never yearned for something larger when definition was superb, the gain would have been mainly aesthetic and he has never felt that anything important was being missed owing to the inadequacy of his equipment.

It is not proposed to enter upon the controversial topic of reflectors versus refractors. If one does not grudge the extra attention required to keep a reflector in perfect adjustment, its performance in revealing planetary detail will equal that of a refractor of the same aperture, particularly if it is mounted with an open, lattice-work tube, when a further improvement may be derived from the employment of an electric fan to keep the column of air above the mirror well mixed. Moreover it has practically negligible chromatic aberration, whereas colour estimates made with a refractor are exceedingly unreliable.

Few amateurs have the facilities for installing even an 8-inch refractor, whereas the tube of the author's 12-inch reflector is only 6 ft. 3 in. long and requires quite a modest revolving shed for housing it in perfect convenience. Mr. F. J. Hargreaves, who has made such an admirable contribution to the Jupiter records, uses a 14½-inch reflector that is no longer and is similarly protected. The author can testify to the somewhat greater satisfaction to be obtained from using the larger instrument.

There is, however, an absolutely essential point connected with the use of a reflector that even now does not seem to be generally appreciated. It is hopeless to expect to see even a fairly good image, if an ordinary Huygenian eye-piece is employed with a short-focus objective; a Ramsden type is a little better but is by no means to be recommended. The reason is that the incomplete spherical correction of common eye-pieces is competent to deal only with pencils of rays that converge at a comparatively small angle, such as those proceeding to a focus from an objective of not less than 12 to 1 focal ratio, whereas the wider angle of the cone reflected from the average mirror requires almost perfect correction of the eye-piece. There can be little doubt that the low opinion held of the defining quality of reflectors forty years ago was mainly due to lack of appreciation of this essential fact. A good orthoscopic or monocentric eye-piece will bring out all the detail in the perfect image formed by a first-class mirror; and although they are more expensive than Huygenians they are essential for even tolerable performance, unless a 'Barlow' lens is used to increase the effective focal length of the mirror. In making measurements with a filar micrometer and a reflecting telescope the fitting of a Barlow lens is almost obligatory, since a positive eye-piece of the Ramsden type is essential. Rather than be continually fitting and removing the micrometer, when a series of latitude measurements was in progress, the author has often used the Barlow–Ramsden combination over long periods with highly satisfactory results; in fact he grew to prefer it to his monocentric, probably because it enabled him to remove the unpleasant effects of atmospheric dispersion, as will be explained later.

Observational Technique

The limitations of visual methods are obvious. Any record of form, intensity or colour, made visually, with no other aid than the telescope itself, is really no more than an expression of the observer's opinion; and the weight that can be given to this opinion depends upon the conspicuousness of the features recorded and upon the recognised skill and judgment of the observer. The acquisition of these qualities is largely a matter of practice and experience, achieved maybe over a period of many years; while hand in hand with these must go the exercise of a ruthless critical faculty. Indeed, so many pitfalls await even the experienced planetary observer, for the avoidance of which intelligence alone is not sufficient, that a fairly detailed account of observational technique and of the dangers of misinterpreting the results of observation will be found in the ensuing pages and wherever it is felt to be desirable throughout the book.

Drawings of Jupiter

Unless there is some urgent need for making a representation of an object before the planet's rotation has carried it too far from the centre of the disk, it is not desirable to begin a drawing of Jupiter until one has been at the telescope for half an hour or so, during which opportunities will have occurred for timing the central meridian passages of objects that will later appear on the preceding side of the picture. A little simple arithmetic will then enable these features to be assigned at the proper time with reasonable accuracy

38

to their true positions; for it is essential that in making a drawing the conspicuous markings should be used as fiducial points, to which the fainter detail can be related. The author confesses that he has always hesitated to embark upon the making of a drawing of Jupiter, unless there has been to hand a fairly recent series of latitude measurements that will ensure his placing at least the four main belts at their correct distances from the centre of the disk. Some draughtsmen have a personal tendency to place objects too far from the centre of a planet, others to concentrate them there; so it is wise to resort to any legitimate means of attaining reasonable accuracy. When the conspicuous markings have been fixed, one may proceed to fill in the fainter and more elusive details, taking care to begin with the preceding half of the disk, since the objects there are receding towards the limb, while there is plenty of time for the following half, where they are coming progressively into view.

Except in the event of some new and important change upon the planet, it would be foolish to commence a drawing unless the air was steady and definition good. Even in really good seeing the finest detail can be caught only at the best moments, which come all too rarely and are generally over in a matter of seconds. The experienced observer will have learnt to make the most of these precious opportunities but will not feel satisfied that he has seen and drawn correctly, until two or three of them have enabled him to confirm his first impressions. He will often find that these need modification.

Surprise and concern are sometimes expressed that when two or more observers have been drawing the planet at the same time their representations show many and occasionally wide differences in detail. Such might be expected among beginners or those who will never learn; but it is by no means uncommon when the draughtsmen have skill and experience. The explanation, in the author's opinion, of the latter anomaly is that, when the good moments have arrived, the observers have not been looking at the same part of the planet. Jupiter's disk does not present a large area under magnifying powers that can be usefully employed and it may seem reasonable to suppose that one can see it all at once; and so one can, but not distinctly. A surprise may await anyone who will try the following simple experiment. Go out and look at the Moon with the naked eye. The whole Moon together with some nearby stars will undoubtedly be seen at the same time. Now concentrate the attention on the bottom edge of the Moon and, without the slightest movement of the eyes (this may require some self-discipline, so instinctive is the habit of keeping the eyes continually scanning the objects in view), imagine you are about to sketch the detail in the top half of the Moon. You cannot make it out at all clearly. Now it requires a magnification of only about 45 diameters to make the disk of Jupiter appear the same size as that of the Moon; so with a power of 225, which is a reasonable one to employ when studying Jupiter, the disk will have five times the Moon's diameter and twenty-five times its area. If anyone who has seen Jupiter with a telescope is inclined to doubt this, he should wait till it is close to the Moon in the sky and then, keeping both eyes open, view simultaneously Jupiter through the telescope with one of them and the Moon with the other unaided.

New recruits to the telescopic study of Jupiter are nearly always would-be artists. There is of course a great deal of personal satisfaction to be derived from having portrayed, as accurately as one's skill allows, the varied details of what is, after all, a very beautiful object. But it must be realised that far too much time can be spent on making full disk drawings; there are generally far more important matters on which to concentrate the attention. Small sketches of interesting local details may be just as valuable and can be quickly executed; and when a particular belt or zone is in an active state, a long 'strip sketch' of the narrow band of latitude affected, which is continually added to as the planet's rotation brings successive longitudes to the central meridian, can be of enormous value. Plate VIII provides an illustration of how a series of such strip sketches can bring out the development and motions of an outstandingly interesting object in the S. Tropical Zone and of a great disturbance in the S.E.B. From four to six disk drawings each apparition should be ample to furnish a record of the general aspect of the planet's surface.

CHAPTER 5

OBSERVATIONS OF COLOUR

In the literature of planetary observation we find so many references to the colours displayed by the surface features that it seems
advisable to include here a short review of chromatic effects in
general. Many over-hasty conclusions have been drawn by observers
who have failed to recognise the importance of the optical phenomena
involved.

Few would deny, of course, that a refractor is an unsuitable
instrument for use in the estimation of colour. The most that can be
expected of it is that it will reveal contrasts. A reflector is very much
better. Apart from the very slight pinkness of the silver film or the
tendency of aluminium to over-emphasise the blue end of the spectrum, the eye-piece, with which must be included Barlow lens and
diagonal prism if either is employed, is the only instrumental source
of spurious colour. Since most modern eye-pieces are sensibly
achromatic, we may safely conclude that there is little to be feared
from instrumental aberration if the telescope is a reflector. There
remains, however, the Earth's atmosphere, which introduces
chromatic effects that at low altitudes are considerable and are
particularly apt to vitiate colour estimates in the case of a belted
planet like Jupiter. The greater density of the air near the ground,
compared with that at higher altitudes, not only brings about the
well-known effects of refraction but also causes the image of a point
source, such as a star, to appear as a short vertical spectrum with
the violet at its upper and the red at its lower end. This is quite
insensible to the unaided eye, except rarely at sunset over a distant,
clear horizon, when the blue-green image of the Sun's upper limb
may linger for a moment after the red has gone, thus giving rise to
the phenomenon known as the 'green flash'. But under telescopic
magnification it is often obtrusive, particularly in the case of a
planet, whose upper limb (lower in the telescopic image) may
frequently be seen edged with blue, while the lower has a reddish
border. For the same reason, when the planet is belted, blue and red
light will spill over from the edges of the bright zones and colour

the margins of the adjacent darker belts. The general effect in the Earth's northern hemisphere is to make the southern edge of a belt look bluer and the northern edge redder than they would appear if the planet were in the zenith.

When terrestrial observers are concentrated in one hemisphere, it is not difficult to see how failure to appreciate the significance of this phenomenon may foster the drawing of questionable conclusions about the reality of seasonal colour changes apparently observed on a planet. For the magnitude of the spurious chromatic effects depends roughly on the meridian altitude of the planet and this in turn depends on its longitude, or position on the ecliptic. Thus the period of the cycle of the colour effects introduced by atmospheric dispersion will be equal to the planet's year, which is of course the period of its seasonal changes, if there are any.

As was implied above, most of the trouble on Jupiter is likely to occur at the limbs and at the edges of the dark belts, but this does not mean that it is negligible elsewhere—see Chapter 7, p. 66. Mars is not a belted planet and its colour problems are no doubt more complex; nevertheless those who would interpret the changes of tint, suggested by some of the observations, as evidence of a seasonal cycle on Mars should not lose sight of the fact that a seasonal effect of some sort, however small, is likely to be superimposed upon any real variations of colour by the Earth's own atmosphere.

In view of what has just been written observers will be interested to note that the unpleasant effects of atmospheric dispersion may be almost entirely removed if a Ramsden eye-piece is used. This was briefly mentioned in the previous chapter and is due to the fact that this type of eye-piece is sensibly achromatic only fairly close to the centre of the field of view. As the image of a planet is moved away from the centre, its more distant edge develops a bluish border, while the nearer shows red. If, therefore, the image in an inverting telescope be placed at the right distance above the centre of the field, atmospheric dispersion may be practically compensated. Supposing, however, that we are interested in the coloration of the edges of one of Jupiter's belts; we must be careful that we do not first compensate away the effect we are seeking and then start looking for it! This can probably be avoided if we match the limbs exactly; but the compensation should be frequently checked, since any irregularity in the drive of the equatorial mounting will allow the image to wander somewhat in the field of view.

Jupiter's surface sometimes abounds in colour contrasts; but nothing even remotely resembling the red of a geranium or the green of the grass is ever seen there. These contrasts often present a picture of great beauty to a sensitive eye; but when we read some of the rather extravagant and, alas, too often conflicting terms that have frequently been used in describing the various tints, we can hardly doubt that the observer's artistic appreciation must have temporarily lulled his scientific judgment. The author once recorded in his notes that a dark projection at the S. edge of the N.E.B. reminded him of a spot of blue-black ink upon pink blotting paper. No doubt the contrast was strong; but there need be little hesitation in accepting the simile as an exaggeration.

Any physicist who wishes to advance a theory that will account for the coloration of Jupiter's surface should seek every opportunity of studying the planet himself, making a sufficient number of observations to become personally familiar with the nature and magnitude of the effects he is trying to explain; it would be difficult to make a just assessment of the value of the observations from a study of the literature alone. The visual appreciation of colour is such an individual affair that it can hardly be expected to furnish scientific data reliable enough to be used with confidence as a satisfactory basis for theoretical investigation.

The author's usual impression on turning his telescope to Jupiter is simply of yellow zones and grey belts. Then, with attention, he may note that some of the darker greys are brownish, while there may be a trace of pink about one or more of the lighter ones or possibly a faint bluishness. Occasionally he has seen in very dark markings at the N. edge of the N.E.B. that intensely deep, almost black redness that is characteristic of a dark clot of dried blood. But these effects do not obtrude themselves; he has to look for them. Sometimes arresting, however, is the outstanding whiteness of a zone compared with the pale yellowness of its neighbours. The S. Tropical Zone has at times been conspicuously white. The Equatorial Zone, on the other hand, may appear decidedly sombre and has been described on occasions as tawny in colour.

The most reliable source of information about Jupiter's colouring must surely lie in the photographs that have been obtained of the planet in light of different wavelengths. Evidence that the Red Spot really is reddish, for example, can be gained from even a cursory glance at Plate I; in the ultra-violet picture it is a conspicuously dark object, while in the infra-red it is practically invisible. It will be noted also that the dark projections at the S. edge of the N.E.B. are more pronounced in the long-wave photographs, the implication being that they have a bluish tint.

For quantitative estimation of the colours thus revealed, accurate information as to the transmission of the filters employed is necessary and it is suggested that a profitable line of research may lie in this direction, especially if many more photographs in the light of different wavelengths can be made available. The fact must not be overlooked, however, that photography can no more counteract the effects of atmospheric dispersion than can the eye of the visual observer.

CHAPTER 6

OBSERVATIONS OF POSITION

In view of what has been written above, some of the visual records of form, intensity and colour, especially those of the more delicate features, may seem to the critical student to rest upon somewhat insecure foundations. On no account must this be taken to imply that they are not to be treated with respect; but probably only a trained visual observer is competent to assess the visual observations of others and to assign just relative weights to the work of known individuals.

Observations of position, however, fall into quite a different category; for the positions of even quite difficult objects can be determined visually by an experienced observer with an accuracy that is entirely adequate to reveal and provide a quantitative record of their astonishing motions in longitude. Only if it is desired to investigate very small changes taking place in a short time—of the order of a week or less—do the errors that are likely to occur become significant. It is surely because they do provide strictly quantitative data, that most of our present knowledge of the behaviour of Jupiter's surface features, which is very considerable and quite remarkable, has arisen from the study of observations of position.

Positions on Jupiter's surface are recorded in latitude and longitude as they are on the Earth; but in actual practice the latitude of a spot is generally given descriptively by reference to its location relative to the centre or edge of one of the belts or zones, of which the latitude has been determined independently by the method about to be described.

The Determination of Latitude

Not every amateur is equipped with the means of making visual measurements of the latitudes of Jupiter's belts; for the employment of a filar * micrometer is essential and this accessory is useless

---

* or some form of 'comparison image' micrometer.

unless the telescope to which it is attached is mounted on a clock-driven equatorial. For the benefit of both those who wish to make micrometer measurements of latitude and those who are interested in the records, a description is given here of a recommended procedure for carrying out the measurements and of the standard method of reduction.

In the focal plane of the micrometer, which must also be the focal plane of the objective, there are generally three fine spiders' webs, two of which are parallel while the third is set at right angles to them. The last is fixed; but in most types both the other webs are movable so that, while they remain parallel, the distance between them is variable. They can be brought into apparent coincidence or crossed over one another so that either web is on the right- or left-hand side. The movement is effected and the distance between the webs measured by a micrometer screw which is usually of 0·01 inches pitch, the head of the screw having a hundred graduations, each of course giving a reading to 0·0001 inch, which by estimation may be extended to 0·00001 inch. The diameter of the image of Jupiter formed in the focal plane of an objective of 10 feet focal length is about 0·03 inch or less, according to the distance of the planet from the Earth.

Calling the movable webs A and B, it is usual to adjust them so that when they coincide A gives a zero reading. B is then left permanently in this position and the measurements are made by moving A alone. There will inevitably be a small error in the zero reading; but this can be eliminated by making half the settings with A on one side of B and half with it on the other, when the error will disappear upon taking the mean. The slides carrying the webs are held by light springs against the threads of the screws and, as the web A is being brought into its final position before a reading is taken, it is important that the last movement should be made against the compression of the spring, in order that backlash in the screw may be avoided. This practice has the additional advantage that in half the settings the final movement separates the webs and in the other half closes them; if therefore an observer has a personal tendency to stop too soon or to proceed too far, this has a good chance of being averaged out.

If it is desired to measure a planet's diameter or the angular separation of the components of a double star, the constant of the micrometer must be known, that is, the number of seconds of arc corresponding to one complete revolution of the screw for the particular telescope upon which it is mounted; but measurements of the latitude of markings do not necessarily require a knowledge of the constant for their reduction, although it can be used to eliminate to a large extent systematic errors in setting on the limbs of the planet.

In practice the two edges of a belt are measured separately when the belt is wide enough to warrant this procedure; but as a rule this is practicable only for the N.E.B. and the S.E.B.; for the others the middle of the belt is the target. The N.T.B. and the S.T.B. are nearly always conspicuous enough for measurement; but further from the equator there is often insufficient contrast for accurate settings to be made.

It is not easy to place the micrometer web with confidence on the edge of a belt; for not only does the web itself hide the detail over which it is placed but diffraction effects at its edges may cause slight apparent displacements. Setting on a limb is probably even more precarious. Irradiation always makes a bright object look larger than it really is when it is seen against a dark background and will tend therefore to make the N. and S. limbs appear too far from the equator by an amount that should be systematic for a given observer. The alternative method of reduction given below is based upon the hope that the error will be the same at both limbs, when it will disappear if the diameter of the planet is obtained from the ephemeris instead of from the micrometer readings.

Let us now suppose that a measurement of the latitude of the N. edge of the N. Equatorial Belt is to be made. The first operation, after ensuring that the image of Jupiter and the webs are simultaneously in focus, is to turn the micrometer head until the two movable webs are parallel to the edge of the belt. The image of Jupiter is then moved in the field of view, by means of the slow-motion controls of the telescope, until the web B is tangential to the north limb, when web A should be placed between B and the edge of the belt to be measured. A is then moved away from B against the compression of the spring until it becomes coincident with the N. edge of the N.E.B.; if it should be moved too far, it must be brought back and moved forward again, not just brought back to coincidence, because of the tendency to backlash. A setting made by separating the webs is conveniently termed 'direct', one made by closing them 'indirect'. This direct operation is repeated at least two more times and the separate readings are recorded. Then the N. edge of the belt is brought into coincidence with web B, while A is moved away from it to tangency with the S. limb. This is done the same number of times as before, three being the author's practice, and the readings taken. After this the indirect settings are made. The edge of the belt is made to coincide with B, while A is crossed over to the other side of it until it lies well outside the N. limb, after which a motion of the screw in the same direction as before will bring it gradually up to tangency. Finally the S. limb is made to touch B, while A is brought up to the N. edge of the belt from the northern side of it. The number of indirect readings should be the same as the number of direct readings. Unless the drive of the telescope is superlatively good, the image of Jupiter may be expected to wander a little after it has been correctly placed with regard to web B. When this happens it is a great waste of time to bring it back by means of the slow-motion controls, in which there is considerable backlash; instead, just press gently in the appropriate direction against the tube of the telescope with one finger of the disengaged hand and with a little practice it will be easy to keep the image in the position required.

The micrometer readings may be conveniently tabulated as in the following example, in which the figures given are those actually obtained by the author in a series of settings on the N. edge of the N.E.B. made on 1943 January 31.

Latitude of N. edge N.E.B.

	From N. limb	From S. limb
	102·7	188·2
Direct	102·5	189·7
	103·7	189·0
	103·9	190·1
Indirect	104·1	193·6
	104·8	190·5
Sum	621·7	1141·1

Before the reduction of the observations is considered it should be noted that for any planet whose shape is an oblate spheroid, the Earth for instance or Jupiter, there are two different ways of reckoning latitude. On the Earth 'geographical' latitude is commonly used, both on maps and by navigators who wish to determine their positions at sea or in the air. It is the angle made between the axis of the Earth and the tangent plane at the point concerned. In many astronomical problems, however, 'geocentric' latitude is more fundamental; this is the angle between the equatorial plane and the line joining the place to the centre of the Earth. If the Earth were a perfect sphere, the two would be identical everywhere; as it is very nearly spherical, the difference is always small, being zero at the poles and at the equator and reaching a maximum of only about $11\frac{1}{2}$ minutes of arc near latitude 45°. On Jupiter the corresponding latitudes are named 'zenographical' and 'zenocentric' from the Ionic genitive of the Greek 'Zeus'. As Jupiter is far more oblate than the Earth, the difference between these may be significant; its greatest value reaches nearly 4°.

For some reason it has been the custom for many years to publish the latitudes of Jupiter's belts as zenographical latitudes. This is rather to be regretted, as zenographical latitude could only be of interest to inhabitants of Jupiter, while zenocentric latitude, being the angle between the radius vector and the equator, conveys more meaning to outsiders like ourselves. However, if a change of practice were to be introduced, it would involve a break in the continuity of the records; so it is probably best to leave matters as they stand, seeing that the conversion of one system to the other is a perfectly simple matter.

In order to reduce the micrometer readings to either zenographical or zenocentric latitude the value of the quantity tabulated as $D_\oplus$ must be taken for the date of measurement from the *Nautical Almanac* or one of the national ephemerides. This is the angular elevation of the Earth above the plane of Jupiter's equator as seen from the centre of the planet and is positive to the north. The reduction is then effected as follows:

If n and s are the measured distances of the belt or one of its edges from the N. and S. limbs respectively, its distance from the centre of the disk divided by the planet's semi-diameter is readily seen to be $\dfrac{s-n}{s+n}$.

Call this quantity $\sin\theta$.

Let $D'_\oplus = 1 \cdot 0714 D_\oplus$ and $\beta' = \theta + D'_\oplus$.

Then the zenocentric latitude β and the zenographical latitude β'' are given by:

$$\tan\beta = 0 \cdot 9333 \tan\beta'$$
$$\tan\beta'' = 1 \cdot 0714 \tan\beta'$$

See Appendix I.

We shall proceed with the reduction of the actual measurements tabulated above for 1943 January 31.

From the *Nautical Almanac* $D_\oplus = +1°\!\cdot\!31$.

First method

$$\sin \theta = \frac{s-n}{s+n} = \frac{1141 \cdot 1 - 621 \cdot 7}{1141 \cdot 1 + 621 \cdot 7} = +\frac{5194}{17628} = +0 \cdot 2946$$

$$\theta = +17°13; \; D'_{\oplus} = +1°40; \; \beta' = +18°53$$

log tan β'	9·5253
log 1·0714	0·0300
log tan β''	9·5553

$\beta'' = +19°8$

Second method

The constant of the micrometer has been determined as being 14″88 and $s-n$ is twice the measured distance d of the edge of the belt from the centre of the disk. In seconds of arc therefore,

$$d = +\frac{5 \cdot 194}{12} \times 14 \cdot 88 = +6 \cdot 44$$

From the *N.A.* the polar semi-diameter is 21″34, giving

$$\sin \theta = +\frac{6 \cdot 44}{21 \cdot 34} = +0 \cdot 3018$$

$$\theta = +17°56; \; \beta' = +18°96$$

log tan β'	9·5360
log 1·0784	0·0300
log tan β''	9·5660

$\beta'' = +20°2$

It will be noted that $n+s$ is the measured polar diameter; so the semi-diameter in seconds of arc, derived from the settings, is given by

$$\frac{17 \cdot 628}{12} \times 14 \cdot 88 = 21 \cdot 86$$

This is about half a second too large, as might be expected from irradiation at the limbs. Thus the second method of reduction, taking the figure from the *N.A.*, should be an improvement on the first, since the published values have been checked by the durations of eclipses of the satellites as they pass through Jupiter's shadow.

The Determination of Longitude

Longitudes on a planet are referred to an arbitrary, standard, zero meridian. On the Earth this meridian passes through the position once occupied by the central web of the old Transit Circle at the Royal Observatory, Greenwich; on Mars it is defined by the tip of a little light wedge-shaped area, known as 'Fastigium Aryn', that separates the two prongs of 'Dawes' Forked Bay'. But on Jupiter no solid surface is visible and it is impossible to choose a standard meridian that can be defined by the ever-changing configurations of the clouds in the planet's atmosphere. We infer from observations of these clouds that the solid body of Jupiter rotates on its axis in just under ten hours; and the only convenient way of selecting a standard meridian is to choose one more or less arbitrarily that rotates in space about the planet's axis in approximately this period. So long as its period is convenient and strictly uniform, its exact value is immaterial.

Early observers * of Jupiter discovered that rotation periods derived from markings near the equator were in general about five minutes shorter than those given by features in temperate and polar latitudes; this difference is great enough to render the use of a single zero meridian quite unsuitable as a reference plane for both classes of objects. Two standard meridians, which with their respective rotation periods are known as System I and System II, are therefore employed, all spots lying within about 10° of the equator being referred to System I and the majority of those beyond these limits to System II. The rotation periods of the standard meridians are in System I $9^h 50^m 30^s.003$ and in System II $9^h 55^m 40^s.632$. Since the choice of period is arbitrary, it may seem curious that the times are not rounded off at least to the nearest second; the reason for this apparent absurdity is that they have been calculated from adopted rotations of exactly $877°.90$ and $870°.27$ in longitude in twenty-four hours for Systems I and II respectively. The latter represents very closely the rotation of the Great Red Spot between the oppositions of 1890 and 1891.

The ephemerides publish the longitudes of the apparent central meridian (C.M.) of Jupiter for some specified hour, usually midnight, for every day when the planet is observable, a separate table being given for each of the two systems. Ancillary tables, such as the comprehensive examples reproduced here as Appendix III, give the change of longitude in any given time; so it is a matter of simple arithmetic to work out the longitude of the C.M. in either system at any moment. If a spot is recorded as having been brought by the planet's rotation to appear on the C.M. at a certain time, then the longitude of the C.M. at that time is the longitude of the spot in the appropriate system. Recording the transit times of as many spots and markings as possible across the central meridian is the first and paramount duty of every serious student of the general behaviour of

* e.g. Cassini, 1690.

Jupiter's surface features. From such records and from such alone has been derived the bulk of what we now know of the amazing longitudinal currents, each confined to and characteristic of its own narrow zone of latitude, that appear to operate, some of them temporarily, others permanently, in the upper strata of Jupiter's atmosphere. During the last sixty years our knowledge of the number, the rates of drift and the general behaviour of these currents has increased quite astonishingly; but there is still a vast quantity of observational work waiting to be done, not only for the sake of the continuity of the record but on the quite considerable chance of discovering new drifts and more particularly of witnessing and recording the recurrence of one that has so far been observed on only one occasion. For instance, will the next conspicuous darkening of the Great Red Spot be associated, as it has been on previous occasions, with a marked lengthening of its rotation period and shall we see again the spots of the 'Circulating Current' of 1932–34 and, if so, will they complete the circuit next time or vanish, as they did before, when about three-quarters of the way round? Again, the extremely rapidly moving spots at the S. edge of the N. Temperate Belt, seen first in 1880 and again in 1891, were dormant, with the exception of a single recorded spot in 1926, until 1929. They were in operation again for two or three years from 1939; when will they appear next and will they have a similar rapid drift? These and many other fascinating questions may be answered by anyone who will record the times of the various spots and markings as they pass across the central meridian. But long hours of painstaking and systematic observation are needed to carry out this programme effectively; Jupiter rarely discloses his secrets to the casual observer!

The method of observation is simplicity itself. Nothing more than the telescope itself is required except a reliable watch; for the times need not be read more accurately than to the nearest minute. It is surprising how rapidly the rotation of the planet causes the apparent displacement of objects near the centre of the disk. If, after noting the position of some spot near the C.M., one retires from the eye-piece for even three minutes, the change on resuming observation is immediately obvious; for in three minutes Jupiter will have turned upon its axis through nearly 2°. When watching is continuous, a marking seldom appears central for longer than two minutes; and by taking the mean of the times when it is first and last judged to be on the C.M. a satisfactory record of its transit should be obtained. Owing to inevitable errors of judgment the times recorded by a reliable observer will often differ by one or two minutes from the true time of meridian passage, seldom by more than three minutes; but even if the error should be as great as five minutes the observation may occasionally be a valuable one. For what we are trying to ascertain is the average rotation period given by the particular spot over a duration of from, say, three weeks up to several months. If there were an error of five minutes in opposite senses in each of two observations of the same spot made thirty days apart, the error in the rotation period, deduced from these two observations alone, would be eight seconds; but normally there would be a number of other observations that would contribute their weight to the mean and thus greatly reduce the effects of the errors. If there were no other observations, then no attempt would be made to publish a period derived from such meagre data, unless the rate of drift should turn out to be exceedingly rapid; in the latter case, provided that identification of the two observations as being of the same object was beyond dispute, an error of a few seconds would be a small price to pay for success in establishing the existence of a current of which the rotation period differed from that of its neighbours by perhaps two minutes. For the last forty years and more the practice of the directors of the Jupiter Section of the British Astronomical Association has been to refrain in general from deriving periods for spots that have not been observed on the central meridian five or six times over a period of at least thirty days, unless special circumstances suggested that something of value might be lost if the observations were excluded. The number of C.M. passages with the interval that elapsed between the first and last has always been published in the tables of rotation periods.

One way in which a poor observation may prove to be of considerable value is in confirming the identity of a feature, in the record of which there is a gap owing to a long spell of cloudy weather or to some other set of circumstances that may have combined to prevent it from being observed on the C.M. for some time. A spot may have been well observed for a month or so at the beginning of its career and another spot for a month or so at the end; their positions and motions suggest that the two are identical, but there is a gap of, say, three weeks without a transit. If during this interval there had been a single record of a similar object in about the right longitude, this might have enabled the identity of the two spots to be established, no matter how unsuitable the intermediate observation might have been for inclusion with the others in the determination of the rotation period. For this reason it is desirable, when beginning a spell at the telescope or when a sudden improvement in the definition reveals detail that has previously been undetected, to make estimates of the probable times of transit of as many features as possible that are already past the C.M. If a spot is judged to have been central less than ten minutes previously, the estimate may be of great value, provided it is recorded as being an estimate only; even up to a quarter of an hour an experienced observeɪ may make a guess that it is worth while to enter as such in his notebook. The same may be said of objects that are approaching the C.M. when the time comes to close down, or when clouds threaten, and of any that may have been inadvertently missed, perhaps while the attention was concentrated upon others or upon making micrometer settings.

When two or three experienced observers have independently recorded the transit times of the same series of spots, it will not be surprising if, on examination, their estimates reveal systematic differences. Observer A may have a personal tendency to record his times a couple of minutes early, while B may be about right and C a minute or so later than B. There is little harm in this, since, when it is known, corrections can be applied; but the differences should be checked occasionally, as they may not remain constant and are particularly liable to variation between one apparition and the next. An observer can obtain a good idea of his own 'personal equation' by noting the times when the satellites and their shadows appear to him to be on the C.M. as they cross the disk. Satellites III and IV always appear as conspicuous dark spots in transit, except when near the limbs; I looks grey and is easy to see only if the background is one of the bright zones, while II is about as bright as a zone and may be quite difficult unless it is projected against a uniformly dark belt. The rate of apparent motion of a satellite, whilst in transit across the disk, is sufficiently uniform for the time of its passage across the C.M. to be taken as the mean of the times of ingress and egress; but although the motion of a shadow is sensibly uniform as seen from the Sun, this is far from being the case when its progress is viewed from the Earth, except within a few days of opposition, owing to the angle between the Sun and the Earth as seen from the centre of Jupiter. However, the time when a shadow will reach the C.M. may be readily calculated; for if T_1 and T_2 are the times of the beginning and ending of the passage of the shadow across the disk, the correction that must be applied to the mean, $(T_1+T_2)/2$, to obtain the true time of its meridian passage as seen from the Earth is given by (T_2-T_1) $(0\cdot5\cos\theta+\tan\theta/2-0\cdot5)$, where $\theta=A_\oplus\sim A_\odot$, $A_\oplus$ and $A_\odot$ being taken from the *Nautical Almanac*, where they will be found in the same table as $D_\oplus$. They are called the 'zenocentric right ascensions of the Earth and Sun' respectively and are measured in the plane of Jupiter's equator from its vernal equinox. The correction is to be added to $(T_1+T_2)/2$ before and subtracted after opposition. See Appendix II.

Having found his personal systematic error, an observer should hesitate before trying to correct it at the telescope. No doubt he would succeed in reducing his mean error, which in any case hardly matters; but so strong is the tendency to 'revert to type' that the 'scatter' in his recorded times might well become greater than before, which is definitely undesirable.

Below are tabulated all the instances during the months of January and February 1932 when Mr. F. J. Hargreaves and the author independently recorded the transit times of the same features. Column 5 gives the differences between their times in minutes in the sense Hargreaves minus Peek. The significance of the figures will be apparent to the reader before he has reached the end of this chapter.

Date	Feature	F.J.H.	B.M.P.	H.−P.
Jan. 21	d. N.E.Bn.	22^h 07^m	22^h 06^m	$+1^m$
	d. N.E.Bs.	22 07	22 08	−1
	d. N.E.Bn.	22 35	(22 35)	
24	w. N.E.Bn.	22 45	(22 48)	
30	f. end STD.	22 20	22 21	−1
	d. N.E.Bs.	22 39	22 35	+4
	w. N.E.Bn.	22 43	22 42	+1
	d. N.E.Bn.	22 57	(22 52)	
	d. N.E.Bs.	23 24	23 23	+1
	d. N.E.Bn.	23 26	23 27	−1
Feb. 17	d. N.E.Bn.	19 45	19 47	−2
	w. S.T.Z.	20 26	20 29	−3
	d. N.E.Bn.	20 30	20 30	0
	w. N.E.Bn.	20 44	20 45	−1
	d. N.E.Bn.	20 58	20 59	−1
	p. end STD.	21 15	21 12	+3
	d. S.T.Bs.	21 30	21 30	0
	d. N.E.Bn.	22 04	22 04	0
	d. N.T.Bn.	22 26	22 26	0
	w. N.E.Bn.	22 26	22 30	−4
18	w. E.Zn.	19 49	(19 55)	
	d. S.E.Bn.	19 51	(19 55)	
	centre RSH.	19 57	19 57	0
	w. S.T.B.	20 37	20 42	−5
	d. S.E.Bn.	20 48	20 51	−3

Figures in brackets are estimates, either of objects not actually seen on the C.M. or of those whose times were regarded for some other reason as approximate only.

It is true that this method has often been criticised; its justification rests upon what it has accomplished. In 1896 A. Stanley Williams published his classical paper, entitled 'On the Drift of the Surface Material of Jupiter in Different Latitudes',* in which he announced the permanent or semi-permanent existence of nine separate surface currents; the rotation periods of the spots moving in each current were characteristic of their latitudes, which in some instances were confined within very narrow limits. Williams' results were based largely upon his own visual transit estimates, although he included the work of a number of other observers; and when in 1901 Rev. T. E. R. Phillips assumed the directorship of the Jupiter Section of the British Astronomical Association, he was already so convinced of the soundness and far-reaching possibilities of the simple procedure advocated by Williams, that the determination of the longitudes of as many spots as possible became and continues to be the most important branch of the activities of this section.

In his Presidential Address to the British Astronomical Association in 1915 Phillips gave details of thirteen well-observed currents; these did not include the rapidly moving spots of 1890 and 1891–92 which are now known to be recurrent. Since then at least five more currents have been added to the list.

One of Williams' most uncompromising opponents was the American astronomer Professor G. W. Hough, who vigorously maintained that actual micrometer measurement was the only justifiable means of determining the longitude of a Jovian marking. This might be true if the ultimate aim of the observation were to establish the accurate position of that particular feature at the time concerned; but we may be hardly interested in the spot as an individual, unless it is one of a very few to exhibit a remarkable motion; even then it is the change of longitude in a matter of weeks or months that is the important consideration and in this a matter of even a degree or two is likely to be trivial. Moreover, the micrometer fails completely in the case of small or delicate markings, some of which are of the highest importance, because of the obliteration of such objects by the web of the instrument.

The author would be the last to claim that simple visual estimates can rival in accuracy the readings of a filar micrometer; but anyone who asserts today that the work of, say, the B.A.A. Jupiter Section would have been more valuable, if all its members had been armed and busy with filar micrometers, has entirely missed another point; which is the time factor. On a really good night features that should

* *M.N.*, Vol. LVI, No. 3.

be recorded may cross the central meridian at the rate of more than twenty per hour. A few of these, though possibly of the highest importance, will be too faint for the micrometer in any case; for most of the remainder there will simply not be time to make the necessary settings. Here is an extract from the author's observing book:

1943 January 30

Projection N.T.Bs.	18^h	46^m
f. end of long d. streak in N.E.Bn.	18	47
Little w. spot in N.Temp.Z.	18	47
Grey condensation on Equator	18	47
D. Streak in N. component S.T.B.	18	48
f. end of long d. projection N.E.Bs.	18	49
Light spot N. of N.E.B.	18	49

It would be instructive to watch (and perhaps to listen to) an observer who was trying to cope with these with the aid of a micrometer, particularly if there were a periodicity in his driving worm.

It may be concluded then that, so long as simple eye estimates continue to produce such striking results, there is no urgency for the development of a more accurate method of observation. Even in the case of the 'oscillating' spot of 1940, the remarkable motion of which will be discussed later in this book, it is doubtful whether the use of a micrometer would have revealed much more than was recorded without its aid. When, however, the need arises for the study of really small irregularities in the motions of individual spots, by all means let the micrometer come into its own.

The Derivation of Rotation Periods

When the transit times have been reduced to longitudes, the results are plotted on squared paper against the dates of the observations. Satisfactory scales have been found to be 20 days and 40° to 1 inch. It is convenient and has been the general practice, at any rate among British observers of Jupiter, to depart from the usual convention of plotting the independent variable, time, horizontally; a diagram with the normal orientation is far more suggestive if rotated clockwise through a right angle. Longitudes then increase to the right, as they do when Jupiter is viewed in the telescope; and as the eye runs down the diagram the history of the positions of the various features unfolds itself as though it were being actually observed.

Separate charts are prepared, or inks of different colours may be used on the same chart, for each latitude in which a well-defined current is known to operate. When longitudes that have been obtained by more than one observer are plotted on the same diagram, it is convenient to employ distinguishing symbols, such as circles, squares and triangles, to denote those due to the different individuals; one advantage of this is that any desirable allowance for systematic error can be made when reading from the chart, instead of by the more laborious process of correcting all the longitudes numerically.

When all the observations of spots in any particular latitude have been plotted, the first task is to make correct identifications of objects that have been recorded more than once. This is a matter partly of experience and partly of common sense. In the absence of any detailed description in the observer's notes, the presence on the chart of a number of points, lying nearly on a straight line or on a not too pronounced curve, will immediately imply that they are all records of the same object; but the number of them, taken in conjunction with their distribution in time, will be the deciding factor. For instance, three records of a spot in approximately the same longitude obtained within seven days would undoubtedly be of the same object; but if, after an interval of thirty days from the last of these, another spot had been plotted and again another after a further twenty days, the identification of both or either of these with the first spot would in general be highly unsafe, even if the agreement in longitude were good; for the first three observations would have been too close together to indicate with any accuracy the rate of drift, and there would be no means of deciding whether or not the agreement of the longitudes had been fortuitous. On the other hand, if the five observations had been fairly evenly distributed in time over a period of thirty days and lay on a reasonable straight line or even a gentle curve, it would indicate almost conclusively that they were of the same object and any displacement in longitude could be taken as real, with a strong inference that any pronounced curvature indicated acceleration. Sometimes the chart will imply that a spot that was previously single split into two, which perhaps coalesced again; but caution must be exercised here, since the spot may have been double all the time but only visible as such under conditions of especially favourable seeing. It is rather surprising how many degrees of longitude may separate the components of a double projection at the edge of a belt, that will appear to be a single feature except when conditions are good.

The identifications having been made, the graphs of the various spots are connected up by lines on the chart and numbered, usually in the order of the longitude, actual or extrapolated, of each spot on the date of opposition.

Everything is now in readiness for finding the mean rotation period of each spot. Even when the graph is sensibly curved there is little to be gained by making a laborious analytical solution. Indeed the only features on which the author has ever employed the method of least squares are the Great Red Spot and the two 'oscillating' spots of 1940 and 1941–42, which will be described later. All that is necessary is to take a piece of black cotton and stretch it along the course of the spot on the chart, varying the gradient until the eye estimates that the best representation of the mean motion has been found. The change of longitude in some convenient multiple of thirty days is then read off and the conversion to rotation period effected by means of tables.

The following Table, prepared by Rev. T. E. R. Phillips, is useful; but if many conversions have to be made, the author recommends

Table for converting Change of Longitude to Rotation Period

Change of Longitude in 30 days	Rotation Period	
	System I	System II
	$9^h \, 50^m \, 30^s\!\cdot\!003 \pm$	$9^h \, 55^m \, 40^s\!\cdot\!632 \pm$
0°·1	0^s·1345	0^s·1369
0·2	0·2690	0·2738
0·3	0·4036	0·4107
0·4	0·5381	0·5476
0·5	0·6726	0·6845
0·6	0·8071	0·8214
0·7	0·9417	0·9583
0·8	1·0762	1·0952
0·9	1·2107	1·2321
1·0	1·3452	1·3689
2·0	2·6905	2·7379
3·0	4·0357	4·1068
4·0	5·3810	5·4758
5·0	6·7262	6·8447
6·0	8·0714	8·2137
7·0	9·4167	9·5827
8·0	10·7620	10·9516
9·0	12·1072	12·3206
10·0	13·4525	13·6895
20·0	26·9050	27·3790

the Critical Table he has computed, from which may be read off at a glance any rotation period to the nearest second from $9^h \, 48^m \, 00^s$ to $10^h \, 00^m \, 00^s$ inclusive and which will be found under Appendix IV.

For changes of longitude within a range of $\pm 100°$ the above table is practically linear; if extrapolated to 180°, the value obtained is numerically about two seconds too small.

It is again perhaps desirable to attempt to disarm the critics; for the introduction of so crude a device as a piece of cotton into astronomical arithmetic would appear to call for some justification. Fortunately the outcome of an event in the author's own experience provides as good a defence of the method as could be desired. During the last two or three years of Phillips' long directorship of the B.A.A. Jupiter Section the author used to assist him in the rather heavy task of preparing the Memoirs for each apparition by plotting all the longitudes and, after the identifications had been agreed by Phillips, by deducing the rotation periods. On one occasion he sent the final results to Phillips, who put them away until they should be required and forgot about them. Some time later Phillips took out the charts and, unaware that the work had already been done, deduced the rotation periods himself, after which the author's reductions were discovered and the two sets of results were compared. Both computers had used cotton; nevertheless, although the periods of at least eighty spots had been derived, two of them differed by two seconds and for the remainder there was either complete agreement or a difference of one second.

CHAPTER 7

THE OBSERVATIONS

General

The reader should now have sufficient acquaintance with the methods by which the more important visual observations of Jupiter's surface features, dating back over the last sixty years, have been made, recorded and reduced, to be able to study the results which are about to be set forth with his sense of judgment alert instead of with the uncritical astonishment, mingled perhaps with a sort of semi-mystical awe, that so many books on the progress of astronomy seem to beget in the minds of the uninitiated.

It is true that much of what we have learnt about Jupiter is astonishing and indeed baffling so far as attempts to afford physical explanations of the phenomena have yet progressed; but it is hoped that the critical reader will at least feel no urge to call in question the reality of the phenomena described. Wherever there is any reasonable doubt, attention will be drawn to it, or the observations will be described in sufficient detail to give the reader the opportunity of forming his own opinion. For the rest, the author can only stake any small reputation he may possess as a student of Jupiter, together with the greater reputations of those who have gone before him, on the essential accuracy of what is chronicled below.

A reminder may not be out of place here that all the phenomena that have been seen, described and measured are atmospheric and relate only to the upper visible surface of the planet's cloud formations. That so much that is at once arresting and puzzling should have been revealed by the study of mere clouds is perhaps not less surprising than some of the phenomena themselves.

It is desirable to consider first one or two aspects of Jupiter's surface as a whole, after which we shall take it piecemeal and discuss each of the belts and zones separately, making subdivisions where they seem to be indicated, as, for example, when the two edges of a belt exhibit different rates of drift of the surface markings. The author had all but written 'surface *material*', an expression that may often be found, possibly even in his own writings; but it must be continually borne in mind that the motion of a marking, though often no doubt identical with that of the matter of which it is composed, may in fact be entirely different. Consider, for instance, the appearance, as seen from the Moon, of a depression approaching the British Isles from the Atlantic. There might be a ring of highly reflecting cloud, surrounding a transparent area in the centre, and the whole formation might be travelling from west to east with some easily measurable velocity; but a lunar observer would need to possess some little insight into the mysteries of terrestrial meteorology before he would realise that the flow of the material was cyclonic and was independent of the motion of the configuration as a whole.

It has sometimes been alleged in the past that Jupiter is like a miniature Sun, in that the rotation period exhibited by its equatorial regions is shorter than that near the poles. The analogy is entirely false. On the Sun the daily displacement ξ of a spot due to rotation can be represented by a simple continuous function of its latitude ϕ, conforming closely to the expression

$$\xi = 14°38 - 2°96 \sin^2 \phi*$$

For Jupiter no such formula can be given. While it is generally true that the equatorial spots have the shortest periods, the most rapidly rotating objects that have ever been observed lay in the northern hemisphere in latitude about $+23°$. In fact, beyond the limits of the great equatorial current, which are approximately $\pm 10°$, the distribution of rotation periods in latitude appears to be almost entirely haphazard and is different in the two hemispheres. Phillips pointed out in 1915 that, omitting the equatorial current, the mean rotation period of the northern hemisphere was decidedly longer than that of the southern, the values being $9^h 55^m 42^s$ and $9^h 55^m 24^s$ respectively; these figures happened to be identical with the means of the north and south polar currents. Since then, as has been mentioned, several other rates of drift have been recorded; but in any case it is difficult to assess the mean period of a hemisphere without an accurate knowledge of the extent of each current in latitude. Some of the drifts, moreover, seem to be actually superposed, for example those controlling the motions of the Red Spot and the South Tropical Disturbance.

The general division of the surface into alternating belts and zones has already been described under 'Nomenclature', where it was noted that the five zones and four belts lying nearest to the equator are permanent features. They are, however, subject to variations of

* H. W. Newton and M. L. Nunn, *M.N.* 111, 413, 1951.

width and intensity and the belts sometimes exhibit two or, occasionally, even three components; it is rather rare, in fact, for the S. Equatorial Belt to appear single, except when the south component has faded almost to invisibility. The whole configuration from the N. Temperate Zone to the S. Temperate Zone is roughly symmetrical with regard to the equator; but beyond these limits there may be little correspondence between the two hemispheres, for even the two polar caps may be very unequal in area.

Not even in the lower latitudes do we find any close similarity when we compare the less regular spots, streaks and shadings that are seen to the north of the equator with those in the southern half of the planet; and never has there been recorded in the northern hemisphere a feature remotely resembling either the Great Red Spot or the South Tropical Disturbance. F. J. Hargreaves has emphasised the fact that different types of markings tend to be characteristic of the latitude in which they appear. To take a few examples, large light oval areas and curving wisps or loops are rarely, if ever, seen anywhere but upon the Equatorial Zone; chains of several little white spots, or light streaks containing tiny bright nuclei, are germane to the middle of the N.E.B.; diffuse dusky patches belong to the Polar Regions, while a certain type of very dark elongated spot and an easily recognised kind of conspicuous round white spot are often to be found at the N. edge of the N.E.B. or in the S. part of the N.Trop.Z.; the white spots often partly encroach upon the N.E.Bn. Now it is true that dark streaks and white spots may sometimes be seen near the N. edge of the S.E.B., for instance; but there is something quite distinctive about the appearance of the former which it would be hard to describe, but which would leave the experienced observer in little doubt as to their environment if it could be arranged for him to view them alone while the rest of the planet was hidden. Many years ago, at a meeting of the British Astronomical Association, a member proposed that these dark oblong N.E.Bn. spots might appropriately be referred to as 'barges', of which they are indeed highly suggestive; whereupon Captain M. A. Ainslie, who was a diligent and devoted observer of the planet, aptly remarked that on no account must we allow canal language to become associated with Jupiter!

Periodicity in Changes of Colour

In 1899 * A. Stanley Williams discussed all the available records of the colours of the two great equatorial belts made between 1836

* *M.N.*, LIX, 7.

and 1898 and found that the maximum intensity of redness in one belt had repeatedly been synchronous with minimum redness, or with a colourless or even bluish phase, in the other, the complete cycle of changes having taken place on the average in a period of 11·08 years. He also found that the colourless state of a belt occurred a short time after the autumnal equinox of the hemisphere in which it was situated; but this cannot be reconciled in the long run with his period, since a sidereal revolution of Jupiter round the Sun occupies 11·86 years. T. E. R. Phillips, in his Presidential Address of 1915 October 27 to the British Astronomical Association, discussed the further observational data obtained since 1898 by members of the B.A.A. Jupiter Section, but found little evidence of periodicity except that a maximum redness of the N.E.B. occurring in 1903 at the same time as a minimum for the S.E.B. was in good agreement with a prediction from Williams' formula. It did appear, however, that an increase of redness in one belt synchronised in most cases with an opposite change in the other, though the variations of the N.E.B. had far the greater amplitude. Phillips suggested that the appearance of the South Tropical Disturbance in 1901 might have had something to do with the break in the definite wave-like changes shown between 1836 and 1898.

The author began an attempt to carry this research further; but at the outset he was faced with so many conflicting estimates of the colours of the belts that it seemed hopeless to anticipate that any conclusion of value would emerge from an investigation that would have involved the expenditure of a great deal of probably fruitless labour. The kind of difficulty with which he was confronted is exemplified by the following quotation from notes made during the apparition of 1916–17 by M. A. Ainslie, who devoted a good deal of time to recording his impressions of Jovian colours:

August 7. 'S. edge of N.E.B. dark grey, then brown coppery tint, shading off through orange to the yellow of N.Tropical Zone.'

October 7. 'Copper colour quite gone from N.E.B.'

October 15. 'Colour not very definite, but there was no red.'

October 20. 'A trace of red in the darker part of N.E.B., otherwise no yellow or brown.'

November 8. 'N.E.B. greyish and ill defined towards S. edge, darker or reddish towards N. edge.'

November 11. 'Strong red tinge in N. edge N.E.B. and greenish to S.'

December 1. 'The pronounced copper red of N.E.B. was very striking and more so when the planet was losing itself in the haze

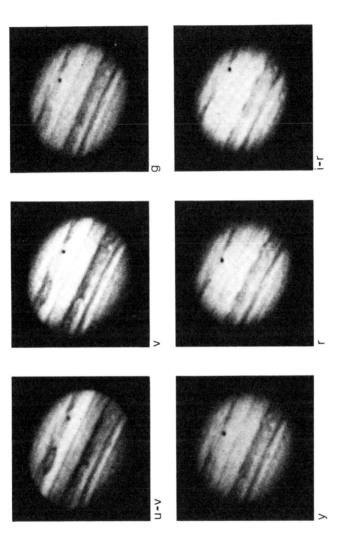

PLATE I.   Jupiter photographed in light of different wavelengths. Ultra-violet, violet, green, yellow, red and infra-red. W. H. Wright, Lick Observatory, 1927 Oct. 2

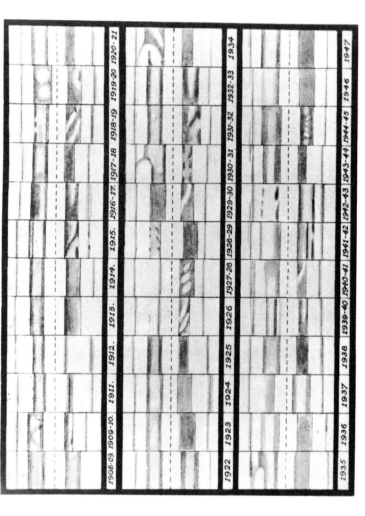

PLATE II.   Changes in the latitudes of the four principal belts, 1908 to 1947

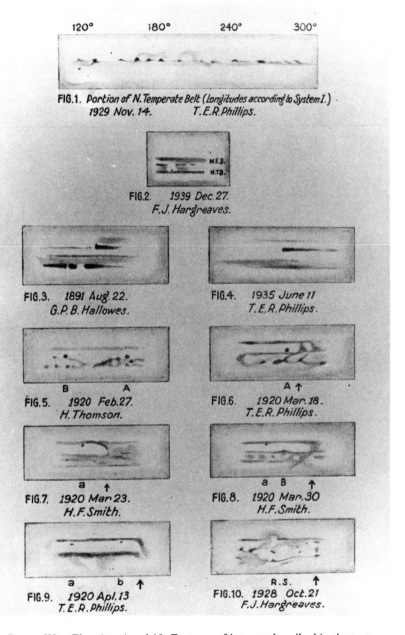

120°   180°   240°   300°

FIG.1. Portion of N. Temperate Belt (longitudes according to System I.)
1929 Nov. 14.   T.E.R.Phillips.

FIG.2.   1939 Dec. 27.
F.J. Hargreaves.

FIG.3.   1891 Aug. 22.
G.P.B. Hallowes.

FIG.4.   1935 June 11
T.E.R. Phillips.

FIG.5.   1920 Feb. 27.
H. Thomson.

FIG.6.   1920 Mar. 18.
T.E.R. Phillips.

FIG.7.   1920 Mar. 23.
H.F. Smith.

FIG.8.   1920 Mar. 30.
H.F. Smith.

FIG.9.   1920 Apl. 13
T.E.R. Phillips.

FIG.10.   1928 Oct. 21
F.J. Hargreaves.

PLATE III. Figs. 1 to 4 and 10. Features of interest described in the text.
Figs. 5 to 9. Progress of the 'circulating' spots of 1920. The arrows indicate
the p. end of the South Tropical Disturbance and the letters R.S. the position of
the Red Spot

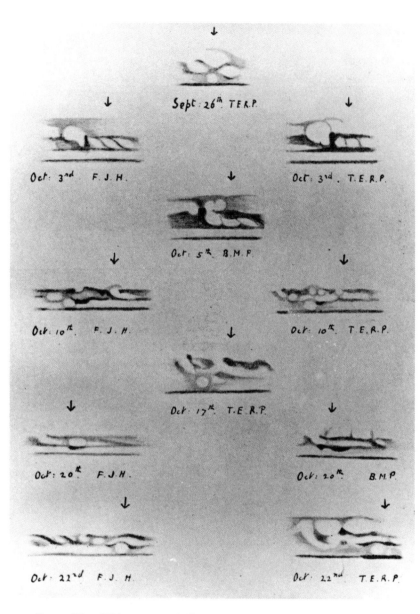

PLATE IV.   White spots and rifts in the North Equatorial Belt in 1927.
F. J. H. Hargreaves   T. E. R. P. Phillips   B. M. P. Peek

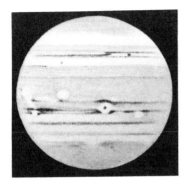

FIG.1.   1941 Dec. 27.
F. J. Hargreaves

FIG.2.   1943 Mar. 2.
B. M. Peek

FIG.3.   1943 Mar. 25.
B. M. Peek

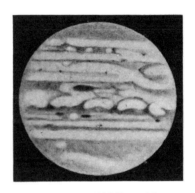

FIG.4.   1928 Sept. 12.
T. E. R. Phillips

FIG.5.   1952 Aug. 31.
W. E. Fox

FIG.6.   1859 Nov. 13.
H. Schwabe

PLATE V

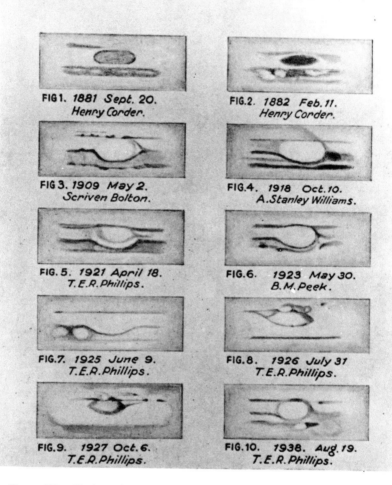

FIG 1. 1881 Sept. 20.
Henry Corder.

FIG. 2. 1882 Feb. 11.
Henry Corder.

FIG 3. 1909 May 2.
Scriven Bolton.

FIG. 4. 1918 Oct. 10.
A. Stanley Williams.

FIG. 5. 1921 April 18.
T. E. R. Phillips.

FIG. 6. 1923 May 30.
B. M. Peek.

FIG. 7. 1925 June 9.
T. E. R. Phillips.

FIG. 8. 1926 July 31
T. E. R. Phillips.

FIG. 9. 1927 Oct. 6.
T. E. R. Phillips.

FIG. 10. 1938. Aug. 19.
T. E. R. Phillips.

PLATE VI. Various characteristic aspects of the Great Red Spot and Hollow

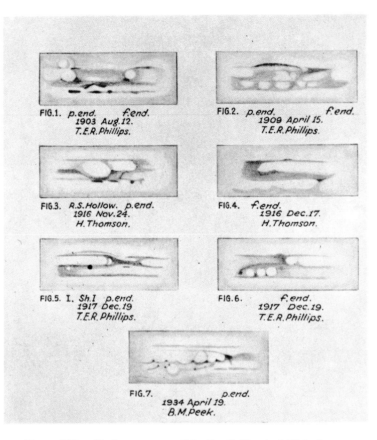

FIG.1. p.end.   f.end.
1903 Aug.12.
T.E.R.Phillips.

FIG.2. p.end.   f.end.
1909 April 15.
T.E.R.Phillips.

FIG.3. R.S.Hollow. p.end.
1916 Nov.24.
H.Thomson.

FIG.4. f.end.
1916 Dec.17.
H.Thomson.

FIG.5. I. Sh.I p.end.
1917 Dec.19
T.E.R.Phillips.

FIG.6.   f.end.
1917 Dec.19.
T.E.R.Phillips.

FIG.7.   p.end.
1934 April 19.
B.M.Peek.

PLATE VII.   Various aspects of the South Tropical Disturbance

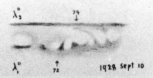

Figure 1. T. E. R. Phillips. 18-in. spec.

Figure 2. T. E. R. Phillips. 18-in. spec. | Figure 3. T. E. R. Phillips. 18-in. spec.

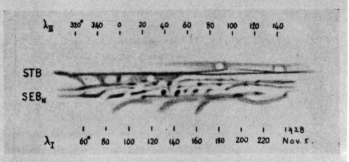

Figure 4. F. J. Hargreaves. 6½-in. spec. ×240.

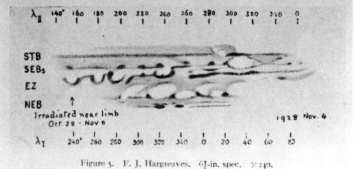

Figure 5. F. J. Hargreaves. 6½-in. spec. ×240.

PLATE VIII.   Strip-sketches of the revival of the
South Equatorial Belt in 1928

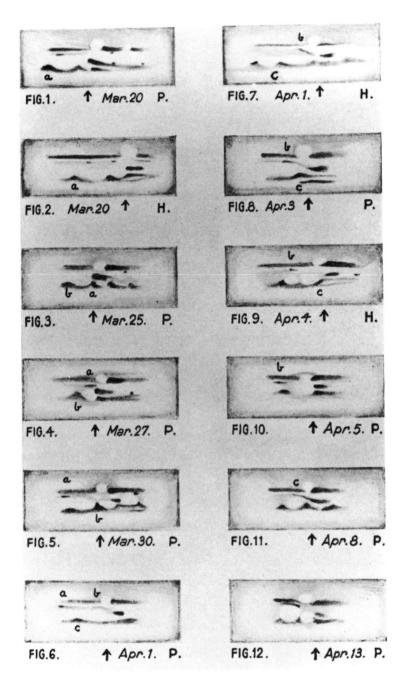

FIG.1.　　↑ *Mar.20*　P.

FIG.7.　*Apr.1.* ↑　　　H.

FIG.2.　*Mar.20* ↑　　H.

FIG.8.　*Apr.3* ↑　　　P.

FIG.3.　　↑ *Mar.25.*　P.

FIG.9.　*Apr.4.* ↑　　H.

FIG.4.　　↑ *Mar.27.*　P.

FIG.10.　　↑ *Apr.5.* P.

FIG.5.　　↑ *Mar.30.*　P.

FIG.11.　　↑ *Apr.8.* P.

FIG.6.　　↑ *Apr.1.* P.

FIG.12.　　↑ *Apr.13.* P.

PLATE IX.　Spots of the Circulating Current encountering the
p. end of the South Tropical Disturbance in 1933,
H. Hargreaves; P. Phillips

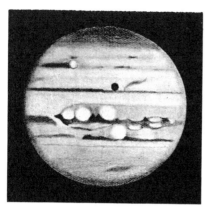

FIG.1. 1926 Sept. 19.
T. E. R. Phillips.
Shadow of Satellite III on disk

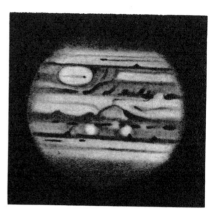

FIG.2. 1929 Nov. 30.
B. M. Peek

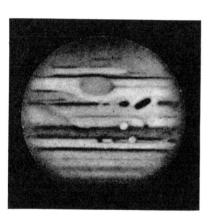

FIG.3. 1943 Mar. 23.
F. J. Hargreaves

FIG.4. 1943 April 4.
F. J. Hargreaves

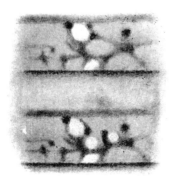

FIG.5. A rapid change observed
by Hargreaves in the South
Equatorial Belt. 1928 Oct. 14.

The lower picture was drawn
only about half and hour after
the upper

PLATE X

PLATE XI.   Jupiter photographed by F. G. Pease with the 100-inch
Hooker Telescope of the Mt. Wilson Observatory

No. 1.   1922 May 29   No. 2.   1921 Mar. 15
No. 3.   1921 Feb. 12   No. 4.   1920 Mar. 28

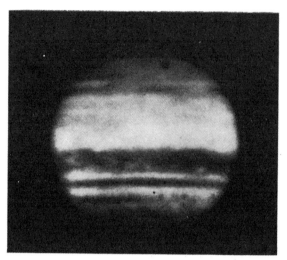

FIG.1.   1950 Oct. 5

FIG.2.   1951 Sept. 25

PLATE XII.   Jupiter photographed in blue light. 200-inch Hale Telescope,
Mt. Wilson and Palomar Observatories

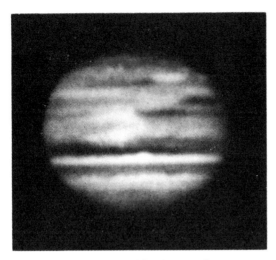

FIG.1.   1941 Oct. 22. 38-cm. refractor.
B. Lyot

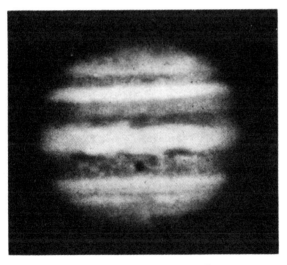

FIG.2.   1945 Apr. 14. 60-cm. refractor.
H. Camichel

PLATE XIII.   'Composite' photographs of Jupiter obtained at the
Pic-du-Midi Observatory

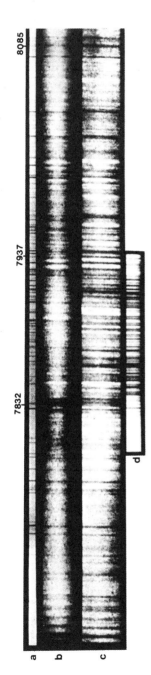

8085   7937   7832

a
b
c
d

Near infra-red spectrum of Saturn and Jupiter

(a) Sun (b) Saturn (c) Jupiter (d) ammonia gas

Fig.1.   The identification of ammonia in Jupiter's atmosphere. The spectra were obtained by
Theodore Dunham, Jr. at the Mt. Wilson Observatory

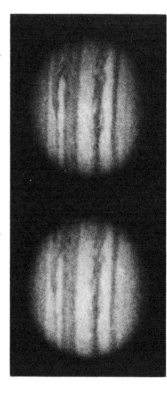

Fig.2.   Jupiter photographed by E. C. Slipher with the 24-inch refractor
of the Lowell Observatory, 1917 Dec. 19

Plate XIV

or emerging therefrom. At times the colour seemed in the half light a pure red.'

December 22. 'Colour of N.E.B. (N. edge) very striking; intense dark red . . . most intense in System II longitude about 30°–60° p. the Red Spot.'

January 1. 'N. edge (of N.E.B.) full orange, very striking and a great contrast with the bluish grey of the N.T.B.'

February 15. 'No red in belts.'

February 21. 'No colour to speak of in the belts.'

March 2. 'N.E.B. a fiery red.'

One more quotation may be of interest; it is taken from the Twenty-first Report of the B.A.A. Jupiter Section, dealing with the apparition of 1918–19:

N. Equatorial Belt. '. . . Observers generally agree in assigning a very warm tone to this belt. It was recorded as brownish red or warm brown by Du Martheray, Bridger, Nangle and Smith; copper colour by Price; ruddy, red, or even brick red in places, by Sargent, Anslie and the Director (Phillips). Williams, who made a long series of colour estimates, also found it very red, the redness attaining a maximum at the end of August (6·5 on his scale on August 28) and then slightly declining. Thomson, however, as a rule, could not see any definite red in it and preferred to describe the colour as khaki, biscuit, yellowish brown, or as resembling that of dead bracken. November 11 and February 18 were the only dates on which he saw a distinct reddish tinge in the belt. Sargent, like Williams, observed that the ruddy tone diminished as the apparition advanced.'

It would appear that the general redness of the N.E.B. may vary greatly even during a single apparition; also that different longitudes may have totally different coloration. There seems, moreover, to be some evidence that the apparent redness of the belts is not independent of the state of transparency of the terrestrial atmosphere, as is implied by Ainslie's note of 1916 December 1; but it is worthy of note that the dates of the maxima and minima of Williams' 11·08-year cycle do not appear to be in the right phase to have been vitiated by atmospheric dispersion.

In 1945 a paper was published by C. F. O. Smith,* in which he presented the results of an investigation he had made into the relative colours of Jupiter's N. and S. Polar Regions from 1891 to 1944. He divided the tints recorded for these areas into two simple categories, warm and cold, and found that over the interval of more than fifty years there was a decided tendency for the S.P.R. to be warmer

* *B.A.A.J.*, 55, 23.

than the N.P.R. There seemed to be grounds here for suspecting atmospheric dispersion, since the majority of the observations had been made from the Earth's northern hemisphere; and when D. P. Bayley, in discussing Smith's paper,* showed that the contrast had been most marked at times when Jupiter lay in high southern declination, the case appeared to be strengthened. It has already been pointed out that this phenomenon can make the north *edge* of a luminous object appear blue and its south *edge* red; but how can it affect the *whole* of an extended area, such as a dusky polar hood? A happy inspiration gave Bayley the solution. The shading of these areas grows progressively darker towards the limbs; so he imagined them as built up from a large number of thin and parallel strips, each strip being slightly darker than the one adjacent to it and farther from the limb. Atmospheric dispersion would give every strip, in both hemispheres, a blue border on its northern and a red border on its southern side; but the effect generated by a lighter strip would predominate over that of its darker neighbour lying on its poleward side. If the width of each strip be now considered as indefinitely small, it will be seen that the integrated result is to make the N.P.R. appear slightly blue and the S.P.R. slightly red. If the reader will examine with a prism one of the illustrations in this book in which strong, progressive shading is shown towards the poles, he should see the effect beautifully demonstrated; by orientating the prism so that its base is first at the top and then at the bottom the tints of the two regions can be interchanged. Thus Bayley, by the ingenious exercise of his imagination, has rendered it highly probable that in general there is little real difference between the coloration of the N. and S. Polar Regions.

Latitudes of the Belts

It is convenient to deal with the latitudes of all the belts together, before going into details concerning the belts and zones individually; the results will then be more readily available for intercomparison than they would be if scattered over the various sections.

The accompanying table is a record of the measurements made with filar micrometers by members of the B.A.A. Jupiter Section from 1908 to 1947. From 1908 up to and including the apparition of 1939–40 they are almost entirely the work of T. E. R. Phillips, but were occasionally supplemented by the measurements of J. E. Phocas, H. Thomson, the author and others; later figures are due to the author.

* *B.A.A.J.*, 55, 116.

Micrometrical Measurements of the Latitudes of the Belts from 1908 to 1947

Appari-tion	N.P.R. S. edge	N.N.T.B.	N.T.B.	N.E.B. N. edge	N.E.B. S. edge	S.E.B. N. edge	S.E.B. S. edge	S.T.B.	S.S.T.B.	S.P.R. N. edge
08–09	+50°2	+36°4		+20°4	+8°8	−7°3	−19°6	−29°8	−41°6	
09–10	49·4	42·4A	+24°5	14·0	8·2	9·0	19·1	29·1	41·7	
11			29·3	+11°9B		7·5	16·4	29·2	44·5	−57°4
12			28·7	+ 8·3B		8·5	17·2	30·1		
13		38·2		20·2	7·9	8·7	19·2	30·4		
14	48·5	37·3	30·4	19·3	8·7	8·9	19·5	29·5	39·0	
15		38·8	26·4	18·0	7·5	8·3	19·8	29·3	43·5	
16–17	48·5	36·9	25·8	17·1	7·8	7·2	19·3	27·5	38·7	
17–18	42·5	38·3	25·7	13·3	6·3	6·9	18·3	26·7	37·0	
18–19		37·4	28·0	18·8	7·0	6·5	17·7	27·8	41·1	
19–20		39·5	29·3	18·7	8·0	4·4	18·1	27·7	42·0	
20–21	45·7	33·1	27·6	12·7	6·1	7·2	19·2	28·5	40·0	
22			31·2	22·0	7·5	6·3	19·2	27·0	39·8	
23			28·8	19·2	6·2	6·6	18·9	29·4		
24			30·6	16·3	4·9	6·1	18·9	28·7	44·2	
25			28·5	13·4	6·1	7·4	18·3	29·5	44·1	
26			25·1	13·9	6·1	(9·5)		28·7	40·7	
27–28		34·7	25·0	17·5	6·2	(8·8)		29·1		
28–29		37·0	25·4	19·8	8·6	8·7	17·2	28·5		
29–30		38·4	25·2	16·6	7·3	5·9	17·6	28·5	41·3	
30–31			26·4	14·7	6·5	6·3	{16·9a / 21·4b}	29·3		
31–32			29·4	21·0	6·3	8·2	19·7	29·7		
32–33			27·5	19·0	6·6	7·5	20·1	29·9	39·8	
34	+40·0C		30·1	14·0	6·1	7·4	{19·1a / 22·6b}	28·6	41·9D	
35		33·1E		18·6	6·4	6·6	{17·4a / 21·6b}	28·7	43·3	
36		35·3	32·2	19·6	7·5	{5·9 / (8·9)}	19·7	28·9		
37		33·8E	27·9	18·3	4·5	(9·3)		30·1	45·4	
38			27·5	15·9	7·4	5·6	18·9	29·6		
39–40		38·9	26·7	15·4	8·7	8·2	16·8	27·2		
40–41			24·8	14·0	7·4	6·4		28·2		
41–42		+36·5	26·8	20·1	7·1	5·6	21·4	29·3		
42–43			27·0	20·2	8·0	7·4		28·5		
43–44			26·4	13·4	7·4	6·1	19·3	28·6	−44·9D	
44–45			29·5	19·6	9·0	5·5	20·3	29·2		
46			29·6	19·9	8·7	7·7		29·2		
47			+30·3	+19·5	+6·8	−7·2	−21·8	−30·5		

Notes on the Table of Latitude Measurements

All latitudes are zenographical.

Figures enclosed in round brackets indicate that the centre of the N. component, not the N. edge, of the S.E.B. is measured.

A. Phillips recorded this as a measurement of the N.N.T.B.; but the high latitude suggests that this belt is missing and that the settings refer more properly to the N.N.N.T.B.

B. The micrometer web was placed on the *middle* of the N.E.B. This belt was narrow and faint in 1912.

C. The object measured was described as 'N.N.N.T.B. or S. edge of N.P.R.'.

D. The belt measured was described as the S.S.S.T.B. It will be noted that in 1934 its latitude was lower than that of the S.S.T.B. in 1935. What is implied is that there was another belt visible between this and the S.T.B. Phillips suggested in the former instance that its designation might have been S.S.T.Bs.

E. The latitude is so low that this may perhaps be considered as a N. component of the N.T.B.

a b } In these years the latitude of the S. edge of the S.E.B. was markedly lower in longitudes preceding the end of the S. Tropical Disturbance than following it. The two figures relate to these different longitudes.

Settings on the Equatorial Band, made in 1938 and 1939–40, reduce to latitudes of +0°1 and −0°6 respectively.

Where no entry is found for a particular belt or component, it does not necessarily mean that the feature concerned was invisible; it may have been too faint or, for some other reason, too difficult to measure.

Plate II exhibits in semi-pictorial form a graph of the two equatorial and the two temperate belts, constructed from the data in the foregoing table. The equator is indicated by a dotted line and continuous lines represent the measured positions. Where a belt is depicted without an accompanying line, it was not measured but has been inserted because drawings of the planet show that it was visible in approximately the position illustrated. The northern and southern limits of the pictures are 0·6 of the polar semi-diameter on either side of the equator and the ordinates plotted are proportional to the sines of the eccentric angles β' which appear in the reductions; the configurations are thus accurately represented as they would have appeared to a terrestrial observer, had the Earth been overhead at Jupiter's equator. North is, as usual, at the bottom.

Students of the planet who wish to look for periodicities in the latitudes of the belts may find Plate II more suggestive than the bare bones of the tabular matter. Probably the most striking feature is the comparative constancy of the latitude of the S.T.B. Phillips in 1926 drew attention to the fact that the breadth of the S.E.B. had been far less variable than that of the N.E.B.; since then, however, it has been somewhat more erratic, though we cannot count the occasions when the S. component has gradually faded (it was entirely absent from the drawings in 1927 and only faintly indicated in 1937) as implying a contraction of the belt as a whole. He also pointed out that there might be a small wave-like recession of the S.T.B. from the equator with maxima in 1913 and 1925. It will be seen that there is another maximum southward displacement shown for 1937; but if this is to be taken as indicating a period coincident with that of Jupiter's revolution, there is a secondary maximum in 1931–32 to be explained. Figures for 1949 are not available; but in 1947 the S.T.B. reached a very high southern latitude. With regard to the northward displacement of the N.T.B., apparently culminating in 1935, attention must be drawn to the fact that an exception has been made for that year in that a measurement described as being of the N.N.T.B. has been plotted; a fainter belt, which was not measured but which appears on the drawings of that apparition and is possibly the true N.T.B., is shown in the diagram, but as the object styled 'N.N.T.B.' falls so well into line with the trend of the N.T.B. in the adjacent apparitions, it was deemed proper to include it.

It should be noted that vernal equinoxes of Jupiter's northern hemisphere occurred in 1914, 1926 and 1938.

In the chapters that follow will be found detailed descriptive accounts of the surface features, mostly grouped into comparatively narrow ranges of latitude, such as are provided naturally by the different belts and zones or by the distinctive atmospheric currents that dominate the regions under discussion.

As it is the accepted custom to publish the numerical values of the different rates of drift in the form of the mean rotation periods of the spots that move with them, values of these periods are given for a large number of years in each section. No claim is made that these data are complete prior to the year 1901, which marked the beginning of Phillips' long and distinguished directorship of the B.A.A. Jupiter Section. The majority of the earlier periods are the published work of individual observers, though Williams, and indeed Phillips until the conclusion of the apparition of 1913, combined whenever possible the periods derived by several observers. Then, in 1914, Phillips adopted the practice, which has been followed ever since by the B.A.A., of combining not the periods but the actual longitudes determined by a number of individuals, plotting them all on the same charts and computing the rotation periods from the aggregate of these determinations. It is clear that more trustworthy results must be the outcome of this procedure, if only because the risk of misidentification of the features plotted can, with care, be made almost negligible. While there is no reason to doubt the reliability of the majority of the earlier publications, the author feels that from 1914 onwards the values of the rotation periods, given in the B.A.A. Jupiter Memoirs, have attained a standard of accuracy that is very high indeed.

We proceed then with our review of Jupiter's surface, beginning with the North Polar Region and working steadily across the planet until we reach finally the South Polar Region. Several features and currents are so striking and so important that, after they have been briefly mentioned in their appropriate settings, complete chapters will be devoted to the detailed discussion of them later.

CHAPTER 8

NORTH POLAR REGION
TO N.N. TEMPERATE BELT

The North Polar Region

The dusky region surrounding Jupiter's north pole extends southwards to about latitude +48° on the average. Its area is very variable and presumably depends on the presence or absence of one or more light zones in high latitude, capable of dividing its southern portion into separate belts. In the last decade of the nineteenth century there were several references to the striated aspect of the region, implying that it was finely banded with a delicacy approaching the limits of resolving power of the instruments employed; but there have been no recent accounts of this phenomenon.

During the latter part of the apparition of 1924 the whole of the surface lying to the north of the N. Temperate Belt was dusky, while from 1938 to 1942 some belts in very high northern latitude were recorded, one of them, in 1939–40, being estimated as having a latitude of 60°.

Reference to the table of latitudes on p. 63 will show that measurements have not been sufficiently regular to suggest any systematic trend in the extent of the area. Apparitions during which the darkness of the N.P.R. has elicited special comment from observers are the following:

1895–96	1924
1902	1930–31
1905–06	1931–32
1912	

It cannot be assumed, however, that these were the only occasions when the shading was more than normally pronounced.

The colour of the region has already been discussed in relation to that of the corresponding region in the southern hemisphere.

As regards surface details the area is commonly quite featureless. Dusky patches or streaks and, less frequently, white spots are sometimes seen, however, and have from time to time provided opportunities of measuring the rotation period associated with spots in high northern latitude. In the following table, which is not necessarily comprehensive, are given a number of rotation periods that have been determined since 1880. They point to a rather steady rate of drift for the features of the N.P.R., with notable exceptions only in 1917–18 and 1930–31. One or two values for spots at the extreme south edge of the area have been omitted, as they were clearly moving in the current that is characteristic of the N.N.N. Temperate spots.

Rotation Periods of North Polar Spots
(North Polar Current)

Apparition	Rotation Period	Apparition	Rotation Period
1881	9^h 55^m 42^s	1916–17	9^h 55^m 47^s
1892	39	1917–18	52
1901	40	1918–19	42
1902	44	1919–20	43
1903	42	1927–28	44
1906–07	41	1929–30	39
1907–08	42	1930–31	34
1908–09	45	1932–33	40
1909–10	45	1939–40	9 55 40
1911	9 55 41		
	Mean	9^h 55^m 42^s	

The N.N.N. Temperate Belt and Zone

The N.N.N. Temperate Zone is by no means a permanent feature of the planet and, when visible, may not extend over the full 360° of longitude; in 1942 January, for example, a following end of the zone was recorded at about System II longitude 15°. The N.N.N.T.Z. is usually featureless; an occasional bright spot has been noted but little more.

The N.N.N. Temperate Belt also is observed only intermittently; it was recorded during about one-half of the apparitions from 1927–28 to 1947 and less frequently in earlier years, probably only because less attention had been given to it. The latitude is generally difficult to measure and seems to be far from constant; as has already been pointed out, it is not always easy to assign the correct designation to high-latitude belts, as for instance in 1938, when fragments of at least three belts could be seen lying to the north of the N.N.T.B.

In 1905–06 there were extensive irregularities in the N.N.N.T.B., while in 1917–18 and in 1929–30 it was fragmentary. In 1939–40 the belt was fairly wide and on one occasion was reported as double by Ainslie; a drawing by Phocas, made at about the same time, shows a bifurcation into two components, starting near System II longitude 85°.

Rotation periods that have been derived from spots, irregularities and discontinuities in the N.N.N.T.B. are given in the following table:

Rotation Periods of Spots in the N.N.N. Temperate Belt
(N.N.N. Temperate Current)

Apparition	Rotation Period
1900	$9^h\ 55^m\ 19^s$
1903	22
1917–18	29
1928–29	20
1929–30	26
1931–32	10
1941–42	19
1942–43	24
1944–45	9 55 10
Mean	$9^h\ 55^m\ 20^s$

An anomalous rotation period of $9^h\ 56^m\ 5^s$, found in 1934 for a single dark spot, described as being on the N.N.N.T.B., will be referred to below in the section dealing with the N.N.T.B.

The N.N. Temperate Belt and Zone

The N.N. Temperate Zone is frequently lost in the general N. Polar shading and, when visible, seldom displays features that call for particular notice. White spots were recorded in the zone in 1903, 1908–09 and 1927–28, while in 1914 attention was drawn to its brightness. During 1940–41 this zone was described as being narrow in some longitudes.

The N.N. Temperate Belt, though often fragmentary or confined to restricted ranges of longitude, has been observed at some time or other during nearly every apparition of recent years, though there does not seem to be any record of its having been seen in 1924.

The vast area of surface lying between and including the N.N.T.B. and the S.S.T.B. is commonly so rich in detail that it may be most difficult to make generalisations about any belt as a whole. The details must be discussed separately or in characteristic groups and, as it is clearly impracticable to review them all, it is the author's duty, by no means an easy one, to select individual objects or classes of object of outstanding significance and to present to the reader the salient facts that are known or may be reasonably inferred about them. But the serious student of Jupiter must realise that, although every effort has been made to avoid the omission of any important or suggestive phenomenon, the details discussed in this book are only a selection; he is strongly advised therefore to consult the original publications. These are, of course, scattered through many different astronomical periodicals; but from 1891 to 1943 the Memoirs of the Jupiter Section of the British Astronomical Association comprise a record that must be unique in its approach to completeness as a report of the progress made in any single branch of research over such a long term of years.

With the reservations of the preceding paragraph in mind the reader may study the following attempt to tabulate a few generalities concerning the aspect of the N.N.T.B. during various apparitions. The columns indicate the dates when the belt was seen double, wholly or in restricted longitudes, and when it was more or less conspicuous than normal. There were frequent occasions when part of the belt was strong and part of it faint; most of these dates have been omitted from the table, which is at best only a partially successful attempt to assess the general quality of the appearance.

General Aspect of the N.N.T.B.

Double	Conspicuous	Faint
1897–98		
		1898–99
		1900
1901	1901	
	1904	
1905–06	1905–06	
	1906–07	
1907–08		
	1908–09	
1911		
	1912	
		1913
1914		
1915		
1916–17		1916–17
1917–18		
1918–19		1918–19
		1923
		1924*
		1925
		1926
		1928–29
		1930–31
		1932–33
		1934
1935	1935	
		1938
		1939–40
1941–42		

* The N.N.T.B. was invisible, or at any rate unrecorded, in 1924.

The delicacy of the doubling of this belt is well illustrated by a quotation from the 32nd Report of the B.A.A. Jupiter Section:

'The N.N.T.B. was often recorded as being exceedingly faint during 1941–42 and was sometimes decidedly inferior to the N.N.N.T.B. There were, however, some darker sections, which in good seeing were generally found to owe most of their strength to very thin, intensely dark lines at the northern edge of the belt. On 1941 December 28 (System II, longitude 90°) the Director (the author) was able to confirm a report he had received from Hargreaves that the belt was double in places; he found that the S. component was rather delicate, and it is possible that the thin dark lines just referred to were in reality portions of the N. component, as this was seen to be very dark in the longitude presented.'

Rotation periods for the N.N. Temperate Current have been derived from dark spots on the N.N.T.B., from the preceding and following ends of darker portions and from light gaps in the belt, with the inclusion of an occasional light spot on the N.N.T.Z. There appears to be a tendency for the light spots on the zone to exhibit a somewhat shorter period than the mean for the belt; but it seems appropriate to include them in the N.N. Temperate Current.

Search has failed to reveal a record, earlier than 1929–30, of any feature resembling even in appearance the most interesting small dark projections that were seen during that apparition at the S. edge of the N.N.T.B. Projections from the edges of some of the belts are common enough and it would probably be correct to state that the S. edge of the N.E.B. is never entirely devoid of them; but there had been, to say the least, a great dearth of them for many years prior to 1929 at the S. edge of the N.N.T.B. In good seeing there was a strong suggestion that some of the spots under discussion were detached and lying in the extreme northern part of the N. Temperate Zone, but if so they were almost in contact with the S. edge of the N.N.T.B. Phillips described them in the Jupiter Memoir, covering the apparition of 1929–30, as having been situated on the zone; but on later occasions they have always been labelled N.N.T.Bs. It was Hargreaves who first drew attention to the unprecedentedly short rotation periods of these objects, which averaged a few seconds less than $9^h 54^m$. No trace remained of them in 1930–31 and they were not recorded again until the apparition of 1940–41, when three similar objects were noted which again exhibited rotation periods comparable with those of the 1929–30 spots. In 1941–42 they appeared in some profusion and again in 1942–43; at least one was recorded in 1943–44 and one, poorly observed but probably authentic, in 1944–45.

Special interest was added to the motion of these remarkable N.N.T.Bs. spots in 1941–42 and 1942–43 by the presence at the S. edge of the belt of an almost precisely similar object, of which the rotation period was nearly one and a half minutes longer than that of the rapidly moving spots and whose continued existence over the two apparitions may be regarded as established. Like the others it appeared to project from the S. edge of the N.N.T.B. and its only distinguishing feature was that it was a trifle more conspicuous. Figure 2 exhibits the longitude chart of markings associated with the N.N.T.B. in 1941–42. The last-mentioned spot is No. 1 on the chart and the rapid relative motions of Nos. 2 to 6 are well shown. The arrow-heads indicate the positions of the preceding and following ends of a darker section of the N.N.T.B. itself, of which the change in System II longitude was small. Since the relative displacements were so rapid, it was obviously not easy to catch one of the other spots in the act of making a passage through No. 1. Apart from the fact that it had to occur at night and when Jupiter was at a reasonable elevation in the sky, clear weather and really good definition were essential within an hour or two of the time of conjunction. On the night of 1942 January 31, however, the author had the opportunity of attempting to observe a conjunction between spots No. 2 and No. 1. Unfortunately haze was rendering the image too dim for really delicate detail to be seen, but during one or two really steady moments the appearance was of a sensibly single object. The presence of two recorded observations on the dotted extension of the track of No. 2 on the chart implies its survival after conjunction; but one could wish that a shorter time had elapsed before it was seen again in order that identity with No. 2 might have been rendered more certain. In 1942–43 one of the spots was well observed four times after conjunction but only once about nineteen days before; two other spots seem to have reached conjunction but were not recorded afterwards.

The anomalous rotation period of $9^h 56^m 3^s$ found for a single dark spot on the N.N.T.B. in 1934 must not be overlooked. In this year, it will be remembered, a solitary dark N.N.N. Temperate spot exhibited a similar period and it seems that for some unexplained reason the most northerly of the North Temperate Currents, which will be discussed in the next chapter and for which this period is about normal, must have extended its influence, as suggested by Phillips, into unprecedentedly high northern latitudes.

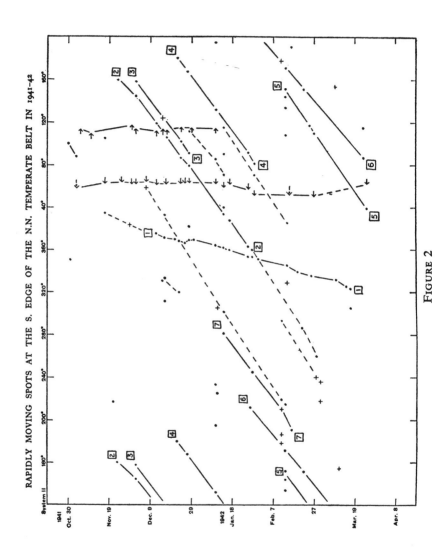

FIGURE 2

The following tables give the rotation periods that can now be associated with the N.N. Temperate regions of the planet.

Rotation Periods of Spots and Markings in the N.N. Temperate Belt and Zone
(N.N. Temperate Current A)

Apparition	Rotation Period	No. of Spots
1888	$9^h\ 55^m\ 41^s$	9
1894–95	39	2
1898	52	2
1899	51	4
1900	37	19
1901	41	31
1902	51	1
1903	41	—
1904–05	48	—
1905–06	49	13
1906–07	42	—
1907–08	44	—
1908–09	42	6
1909–10	38	—
1914	37	1
1915	36	6
1916–17	43	4
1917–18	37	5
1918–19	41	3
1920–21	36	1
1925	46	1
1926	48	1
1927–28	43	2
1928–29	36	6
1929–30	41	6
1930–31	46	1
1932–33	50	1
1934	(63)	1
1935	35	1
1938	47	1
1939–40	35	12
1940–41	35	2
1941–42	39	5
1942–43	38	4
1943–44	38	4
1944–45	(04)	1
1946	9 55 53	2
	Mean $9^h\ 55^m\ 42^s$	

The two figures in brackets have not been included in the mean. That for 1934 refers to the anomalous spot already mentioned. The single object of 1944–45 was the preceding end of a darker portion of the N.N.T.B., similar to many that have been previously recorded; but its unprecedented rotation period indicates that it must have been moving under some influence that has never before been associated with this latitude.

*Rotation Periods of Dark Spots and Projections at the South
Edge of the N.N. Temperate Belt
(N.N. Temperate Current B)*

Apparition	Rotation Period	No. of Spots
1929–30	9^h 53^m 54^s	7
1940–41	54 02	3
1941–42	53 55	6
1942–43	52	6
1943–44	53	1
1944–45	(9 53 51)	1
	Mean 9^h 53^m 55^s	

The dark spot referred to above as No. 1 had a rotation period of 9^h 55^m 19^s in both 1941–42 and 1942–43. The period of a similar feature of the 1944–45 apparition was 9^h 55^m 1^s; but extrapolation of their motions shows that it is very unlikely that the two were identical. Neither has been included in the derivation of the rotation periods for this current. The period of the latter is very close to the abnormal one shown in the previous table for the p. end of the darker section of the belt during the same apparition. It provided a barrier to the passage of the spot, whose period is given in brackets and which has not been included in the mean because it was observed three times only in ten days before the two should have been in conjunction and was not recorded afterwards.

CHAPTER 9

THE NORTH TEMPERATE
BELT AND ZONE

The N. Temperate Zone varies greatly in both width and brightness. Sometimes it is one of the brightest features of the disk, while
on other occasions, as in 1914, it presents a sombre duskiness that
provides little contrast with the belts on either side of it. Its breadth
is naturally related to that of the N.T.B. and depends, in part at least,
upon the visibility or otherwise and upon the latitude of the N.
component of that belt. White spots sometimes appear on the zone
and in 1927–28 there was a fairly extensive disturbed region, the
features of which were found to be drifting at the slow rate usually
associated with this latitude and with the N. edge of the N.T.B.

The N. Temperate Belt also exhibits pronounced variations of
width and intensity. As we are now approaching the more active
latitudes of Jupiter's surface, it is becoming increasingly difficult
to make generalisations about the belts or zones, particularly the
belts, as a whole. Some longitudes may be greatly disturbed, while
others at the same time are inactive; moreover changes may take
place in a week or two and in some cases even in a few days. No
attempt is being made, therefore, to tabulate the apparitions when
the N.T.B. was double, broad or dark, as some of the entries would
have little meaning and might be misleading; on the other hand,
to give a systematic account of all the details that have been
chronicled would be beyond the scope of this book. The student
who wishes to delve more deeply into minutiae has, therefore, no
choice but to consult the original publications. To give here a
satisfactory account of the outstanding phenomena will occupy
space enough.

There was a remarkable sequence of years, from 1904 to 1914
inclusive, when the N.T.B., after being strong in 1903, faded and
remained quite inconspicuous; it is doubtful indeed whether the
true N.T.B. was identified at all during the apparitions of 1908–09,
1911, 1912 or 1914, but by 1915 a revival had set in. In 1924 the belt
was astonishingly broad and, although it is shown single in his

drawings of that year, Phillips estimated that it covered about 9° of latitude.

The belt frequently displays two components, either or both of which may be dark or faint. Sometimes, when a more conspicuous section of the belt is seen, an improvement in the definition will show that the darkening is due to one component only. It is not always easy to decide whether a grey line lying to the north of the N.T.B. should properly be designated 'N.T.Bn.' or 'N.N.T.B.' In 1937, for example, there was a region where the two components diverged until they became quite widely separated and on August 5 of that year Phillips wrote: 'The N. comp. of N.T.B. now (System II longitude 120°) runs very far N. and becomes (?) the N.N.T.B.' The criterion for correct nomenclature would seem to be the rotation periods of the spots on the belt or component; but on that occasion no suitable features were available on the section that ran into higher latitude.

The North Temperate Belt, though usually far inferior in width to the equatorial belts, is remarkable for the disparity between the rotation periods exhibited by spots at its north and south edges. The former almost invariably move at a rate that provides the longest period of all the currents which are recognised today as permanent, while sporadic outbreaks of the latter have given the shortest periods ever observed on the planet. There is, moreover, a quite definite, intermediate rate of drift, associated with other objects in the belt and in 1942–43 all three of them were in operation simultaneously.

The features that move with the longest period, in the drift hitherto known as the N. Temperate Current but here designated 'N. Temperate Current A', are spots, mostly dark and often decidedly elongated in the east–west direction, and longer dark streaks. They are invariably situated either at the north edge of the belt or on the N. component when it is visible. When the N. component is fragmentary, the preceding and following ends of the visible portions will be found to move with this current, of which the average rotation period is a few seconds longer than $9^h 56^m$.

Description of the markings for which the intermediate period has been found is deferred until the rapidly moving spots at the S. edge of the belt have been discussed, as they seem to be not entirely independent of one another.

In 1880 there appeared for the first time in the recorded history of the planet an outbreak of dark spots at the S. edge of the N.T.B. that displayed the shortest rotation periods that had ever been observed on Jupiter, the mean for all the spots having been close to

$9^h 48^m$. The disturbed region was originally not greatly extended in longitude. On October 29 of that year W. F. Denning noted that it occupied less than an hour in passing the central meridian; on November 8, however, he found that it took nearly two hours to go by and on November 23 its passage lasted for $3^h 19^m$, so that it then covered more than 120° of longitude. There is no record that any remnant of the disturbance survived until the following apparition.

The next eruption of dark spots and protuberances at the S. edge of the N.T.B. occurred in 1891, when estimates of $9^h 49\frac{1}{2}^m$ and $9^h 49^m1$ were given for the mean rotation period. On this occasion there were considerable differences between the rates of drift of the individual spots, which in more than one instance were subject to sudden changes of velocity.

Nothing further was learnt about these remarkable phenomena connected with the N.T.Bs. until the apparition of 1929–30, when A. S. Williams was the first to draw attention to the beginning of a third outbreak. The records of Williams' assiduous observations of Jupiter date back to 1877 and he himself witnessed the N.T.Bs. eruptions of 1880 and 1891–92. It is a great tribute, not only to the unimpared acuity of his vision, but also to his ability to appreciate so quickly the significance of the rather delicate objects he detected that, aided by a reflector of only $6\frac{1}{2}$ inches aperture, he was able to forestall other skilled and better equipped observers in announcing a recurrence of what he had last beheld not less than thirty-seven years previously. Here is a quotation from his note-book, dated 1929 October 10–11: 'The appearance of the N.T.B. has so closely resembled (on a smaller scale) that presented during the great disturbance of 1891–92 that it seemed desirable to ascertain as soon as possible if the dark spots on the southern edge of the belt did not rotate at, very approximately, the same rate as that of System I, instead of that of System II. If the former, the spot 3 of Oct. 8–9 should have been in mid-transit at about $4^h 42^m$ on Oct. 11. The night was almost hopelessly cloudy, but the spot was seen distinctly several times during instants of partial clearance, and its transit time estimated at $4^h 36^m$. Rotation period $9^h 48^m8$ very approximate.'

It will be noted that Williams referred to the phenomena of 1929 as being 'on a smaller scale' than those of 1891; but the words were written during the initial stages and a glance at the strip-sketch of 1929 November 14, executed by Phillips and showing about three-quarters of the disturbed region in a more fully developed state—Plate III, Fig. 1—makes it clear that for this region of the planet the spots and protuberances were very conspicuous. The activity reached a maximum probably about the middle of December, after which the number of visible spots began to diminish. Four or five similar objects were followed during the next apparition and they gave a mean rotation period actually a little shorter than that of the 1929–30 spots, which was $9^h 49^m 17^s$ derived from twenty-one spots. Every one of the few spots of 1930–31 had a period within one second of $9^h 49^m 10^s$. In 1931–32 there were still some survivors, which will be referred to again below, as their rotation periods were so much longer that they really fall into the intermediate category.

Those who wish to consult the original publications may refer to the author's paper, *M.N.*, XCI, 941, 1931, and to the twenty-eighth Report of the B.A.A. Jupiter Section, *B.A.A. Memoirs*, Vol. 32, Part 4. The latter was compiled in 1937 by the Director of the Section, Phillips, in collaboration with the author. Comparison of the spot charts accompanying these two publications reveals some discrepancies. The inclusion of a few more observations in the later chart led to a revision of the identification of some of the spots, with the result that the general tendency of the tracks on the earlier chart to indicate a lengthening of the mean rotation period during the latter part of the apparition virtually disappeared. In the author's opinion the diagram in the Memoir is the more trustworthy of the two.

In both charts two facts are well illustrated, namely that the beginning of the disturbance was not confined, as in 1880, to a narrow range of longitude with rapid subsequent expansion and that, from the middle of October 1929 until the end of the apparition, about 240° of longitude, or two-thirds of the circumference of the narrow zone affected, was occupied by the disturbance, while the remaining 120° was practically free from spots. The active region made a complete circuit of the planet, relative to System II, in thirty-eight days.

Almost exactly ten years after the beginning of the 1929–30 outburst the initial stages of yet another eruption were detected. The first dark spots were noted early in October 1939—one had actually been observed on September 23—and by the end of the month there was a fully developed group of them, covering about 180° of longitude. The remaining half of the N.T.Bs. remained for a time unaffected and no spot was recorded there before November 9. On that date the observation of four new spots in the hitherto inactive longitudes extended the length of the whole disturbance by about 80° and the appearance of two or three more objects during the next few days increased it to 320° by the end of November. Owing to the fact that the leading spots had shorter rotation periods than the following ones, the barren stretch of 40° was soon demolished and by the beginning of 1940 the encirclement of the zone was complete. The mean rotation period of all the well-observed spots was $9^h 48^m 57^s$. This outburst persisted throughout the apparition of 1940–41 with a mean period of $9^h 49^m 11^s$; but in 1941–42 an insufficient number of N.T.Bs. objects was recorded to give any definite rotation period, although the 'intermediate' rate was exhibited by another part of the N.T.B., as described below.

In a paper published in 1898, *M.N.*, 59, 76, Denning suggested a possible periodicity in the outbreaks of rapidly moving spots at the S. edge of the N.T.B. It is, of course, possible to fit, without excessive elasticity, a period of roughly ten years to agree with the eruptions of 1880, 1891, 1929–30 and 1939–40; but it is quite certain that the major phenomena were not repeated during any of the apparitions between 1892 and 1929–30, for it is unthinkable that such an occurrence could have passed unnoticed during the years when Phillips was systematically observing Jupiter. It is conceivable, however, that a small outburst may have occurred during the months when the planet was too close to the Sun for observation; but if so, it must have been on quite a minor scale, for otherwise some trace of it would have survived into the succeeding apparition. The return of the rapidly moving spots in 1929 did, however, prompt Phillips to make a search through the records of the years intervening since 1892, in order to discover whether anything of significance had been overlooked. He found that very few dark spots had at any time been visible at the S. edge of the N.T.B.; but re-examination of the longitude chart of the region for the apparition of 1926 revealed a solitary object that had been moving with a rotation period of $9^h 49^m 3^s$. His drawing of the planet, made on 1926 September 19, appears to indicate that it was not quite the only feature of its kind.

Turning now to the markings that have moved at the intermediate rate of drift, which has exhibited a mean rotation period of $9^h 53^m 17^s$ and for which details are tabulated below under 'N. Temperate Current B.', it is to be noted that on most occasions the features that have moved with this current have been preceding ends of darker portions of the belt itself. In 1928–29 there were two p. ends and a dark spot in the belt; in 1931–32 there was a p. end together with four dark and two light spots, the last six still being at the S. edge of the belt, all of which were in the intermediate current; in 1932–33 one p. end had a period of $9^h 53^m 46^s$, while the following ends of two dark portions gave $9^h 55^m 32^s$ and $9^h 55^m 41^s$, which are most unusual figures to be associated with the N.T.B.; in 1942–43 one p. end and one f. end both gave the intermediate period, as did also one p. end in 1943–44. It seems, therefore, as though the intermediate rate of drift appears as the aftermath of an outbreak of the rapidly moving N.T.Bs. spots; the exception that in 1928–29 it preceded by one year the outbreak of 1929–30 may be only apparent, since it followed by only just over two years the minor eruption of 1926.

During the apparition of 1939–40 some of the white spots usually to be found in the N. Tropical Zone were large enough to fill the

Rotation Periods of N. Temperate Currents A, B and C

Year	A Rotation Period	A No. of Spots	B Rotation Period	B No. of Spots	C Rotation Period	C No. of Spots
	h m s		h m s		h m s	
1880					9 48	—
1891					9 49+	—
1892					9 49+	—
1901	9 55 56	2				
1902	9 56 11	1				
1903	9 56 2	11				
1915	9 56 2	1				
1916–17	9 55 58	1				
1922	9 55 56	2				
1926	9 56 2	3			9 49 3	1
1927–28	9 56 6	4				
1928–29			9 53 3	3		
1929–30	9 55 58	4			9 49 17	21
1930–31	9 56 10	2			9 49 10	5
1931–32	9 55 58	5	9 52 46	7		
1932–33	9 56 16	2	9 53 46	1		
1934	9 56 2	5				
1936	9 56 11	1				
1937	9 56 7	1				
1938	9 56 6	7				
1939–40	9 56 0	3			9 48 57	11
1940–41	9 56 3	1			9 49 11	13
1941–42	9 56 1	1	9 52 58	1		
1942–43	9 56 1	3	9 53 29	2	9 49 6	2
1943–44	9 56 27	1	9 53 38	1		
1946	9 56 14	4				
1947	9 56 4	1				
1948	9 56 6	5				
Means	9^h 56^m 5^s		9^h 53^m 17^s		9^h 49^m 7^s (1926 to 1943)	

whole width of the zone; and since many of the rapidly moving N.T.Bs. spots projected southwards from the edge of the belt, these encroached upon the latitude occupied by the northern portions of the white ones. The relative velocity of the groups was so high that it was most difficult to catch a pair actually in conjunction; but such an observation, if it could be obtained, would clearly be of importance. By good fortune Ainslie, Hargreaves and Phillips were all able to observe a conjunction that occurred on 1939 December 27 and all three agreed that the dark N.T.Bs. spot was projected upon the northern part of the white N. Tropical spot. Plate III, Fig. 2 represents Hargreaves' drawing of the configuration. The question of the relative heights of the Jovian markings will be referred to in a later chapter; for the moment this interesting phenomenon is to be noted simply as a record of the independent observations of three skilled observers.

The table on page 84 gives the rotation periods that have been determined for the three currents associated with the N. Temperate Belt and the southern part of the N. Temperate Zone:

We may pause here to contemplate the remarkable juxtaposition of the atmospheric drifts in Jupiter's N. Temperate and N.N. Temperate latitudes. As has already been mentioned in the case of the white N. Tropical spots of 1939–40, the N. Tropical Current frequently extends northwards nearly to the S. edge of the N.T.B., so that its northern limit lies within the zone we are considering. No fewer than six recognised and well-defined currents are therefore associated with the region and in 1942–43 they were all in evidence simultaneously.

The following are the six rotation periods determined for 1942–43 together with the approximate latitudes in which they were found. The latitude of the N.T.B. was actually measured; as, however, no record for that apparition is available for the N.N.T.B., its mean value has been assumed.

Situation of Current	Approximate Latitude	Rotation Period
N.N.T.B.	+35°	$9^h\ 55^m\ 38^s$
N.N.T.Bs.	+33	9 53 52
N.T.Bn.	+29	9 56 1
N.T.B.	+27	9 53 29
N.T.Bs.	+25	9 49 6
N.Trop.Z(n).	+24	9 55 30

It will be seen that it is necessary to traverse only about 11° of latitude for all six currents to be encountered in succession. On the

Earth, to scale, this is equivalent to the difference of latitude between London and Madrid; but Jupiter is so much larger than the Earth, that the distance corresponding there to an arc of 11° is some 13,500 kilometres or 8,500 miles, which is roughly the distance by air to Cape Horn from the nearest point on the Arctic Circle. We see therefore that, although the difference between the velocities of the fastest and the slowest of these currents is 130 metres per second or 290 miles per hour, what appears at first sight to be a most remarkable proximity is not quite so surprising when we realise the true distances involved, particularly when the shallowness of any atmosphere, reckoned in planetary dimensions, is considered. The real miracle is that these drifts appear so consistently in their own narrow zones of latitude, relative to the dusky belts that are displayed in this limited area of the planet's surface. Notice the absence of periods lying between $9^h 49^m 30^s$ and $9^h 53^m 0^s$; also between $9^h 54^m 0^s$ and $9^h 55^m 0^s$. No example of such a rate was recorded, even for an individual spot.

Before closing this section on the North Temperate Belt a brief reference must be made to observations of its colour. Conspicuous hues have seldom been attributed to the N.T.B. Occasionally it has been called brown or purplish; but a large majority of the records describe it as either grey, bluish or blue-grey.

CHAPTER 10

THE NORTH TROPICAL ZONE
AND THE NORTH EQUATORIAL BELT

The North Tropical Zone

This is nearly always a conspicuous zone. Its breadth varies greatly, however, and is complementary to that of the N. Equatorial Belt, of which the northern edge is subject to wide variations of latitude. During one apparition a series of spots in the southern half of the N.Trop.Z. may lie entirely in the zone, while in the next apparition a similar group in the same latitude may appear as features of the N.E.Bn.

Dark spots and streaks, which are characteristic features of the southern part of this zone, and of the N.E.Bn., are rarely observed in its northern half, where as a rule only occasional white spots are seen or the northern portions of those that are large enough to fill the whole width of the zone. As the great N. Tropical Current extends its influence well into the N.E.B., in or close to which the majority of the markings that move with it are situated, it is convenient to treat all the N. Tropical markings together in a discussion of their motions, which is therefore deferred to the next section.

The brightness of the N. Tropical Zone is subject to considerable variation and may be far from uniform; in the latter case estimates of it will naturally depend on the longitude presented. This zone has sometimes been recorded as the brightest part of the disk, as in 1909–10, during the early part of 1911, in 1920–21 and in 1939–40; it was also very bright in 1925, 1935 and 1937 and decidedly bright in 1926 and 1938; in 1914, on the other hand, it was heavily shaded and its sombre hue has been noted on other occasions.

A peculiar and interesting feature of the apparition of 1942–43 was the presence in some longitudes of a narrow and rather faint grey line, which ran for some distance almost exactly along the middle of the zone and which certainly could not be classed as a component of either the N.E.B. or the N.T.B.

The North Equatorial Belt

During the first half of the twentieth century, at any rate, the N. Equatorial Belt has been the most consistently active region of the planet. It is almost certain that, during a single hour of observation, at least one dark projection from its southern edge will be recorded on the central meridian. Such projections are characteristic features and often appear in great profusion; they move with the N. Equatorial Current and the apparition of 1912 was the only fairly modern one, during which rotation periods were not derived for at least a few N. Equatorial spots. In 1911 the S. component of the belt, though dark and rather broad, had been comparatively featureless; nevertheless, the drawings of at least one observer show well-marked projections at its S. edge. Between 1911 and 1912, however, the whole of the N.E.B. faded almost to insignificance; and although a revival set in and projections at the S. edge of the belt were seen again before the end of the apparition, it was too late for a sufficient number of observations to be obtained to furnish rotation periods that could be regarded as reliable. This was most unfortunate; for valuable information might have been gathered as to the manner in which the belt was restored to its normal appearance and could have provided an interesting comparison with the great revivals of the S.E.B. in 1919–20, 1928–29 and 1942–43, which are to be described later. During 1904–05 and 1905–06 the N.E.B. had also been very faint and narrow; and on this occasion too the restoration, which began in April 1906, came so near the end of the apparition that little could be learnt from it. When observation was resumed in August of that year, the N.E.B. had become the broadest and darkest belt on the planet.

The breadth of the N.E.B., though usually exceeding that of either of the Temperate belts, is subject to wide variations, as may be seen from the table of latitudes on p. 67 and from the diagrams of Plate II. The supplementary figures given below for its breadth in degrees of latitude from 1891 to 1907–08 are taken from a table published by Phillips in 1916.

Apparition	Breadth of N.E.B.	Apparition	Breadth of N.E.B.
1891	7°	1900	11°
1892	4	1901	1
1893–94	10	1902	13
1894–95	12	1903	6
1895–96	1·5	1904–05	1
1896–97	15	1905–06	0·5
1897–98	3	1906–07	15
1899	11	1907–08	13

During a large majority of modern apparitions the N.E.B. has been easily the most conspicuous, as well as the most active, belt on the planet, its great width or darkness, or a combination of the two, usually rendering it the first feature to be discerned on the disk with a very small telescope. This is not always the case, however, and its chief rival, the S.E.B., which is generally at least its equal and often its superior in breadth, has on occasions been also the darker of the two.

The belt may appear as a single entity or it may show two definite components; and it is by no means uncommon for the two aspects to be presented simultaneously in different longitudes. Where two components are visible, however, they are seldom so clear-cut or separated by such a definite light zone as are frequently the components of the S.E.B., which may have the appearance of being two quite independent narrow bands. Occasionally, as during the apparitions of 1906–07, 1907–08 and 1941–42, a third component has been visible in some longitudes as a narrow but quite prominent dusky line, running parallel to and in between the other two—Plate V, Fig. 1.

The activity of the northern region of the N.E.B. is hardly less permanent than that of its southern edge, as reference to the table of rotation periods of the N. Tropical current, in which all N.E.Bn. features are included, will show. Certain definite types of marking can be associated with these latitudes. While small, round, dark spots are sometimes seen, the general tendency is for the darker objects to be drawn out in the east–west direction, sometimes into quite long streaks. The lighter spots, when not merely gaps in the N. component of the belt, are usually round, rather large and bright enough to stand out even against the general yellow-whiteness of the N. Tropical Zone. Characteristic of the former are very dark, almost black, elongated objects, very conspicuous and extending over 9° to 15° of longitude and perhaps $2\frac{1}{2}$° to 3° of latitude. Sometimes these occur in pairs, when they may be separated by one of the white spots. The author knows of no attempt to measure the individual latitude of any of these spots; but their placing on drawings of the planet favours the inference that the latitude in which they appear varies from apparition to apparition, though not so widely as does the latitude of the N. edge of the N.E.B. When the N. component is visible, the dark spots are generally threaded along it; but when the belt is wide and single, they appear to be wholly embedded in it, as do also many of the white spots. At other times the dark spots may be at or clear of the N. edge of the belt, while some of the white ones still overlie or indent its northern confines. When we remember that the latitude of the N.E.Bn. has a range of some 9°, these appearances can hardly be reconciled with constancy of latitude of the spots under discussion.

Now and then the appearance of one or more of the N.E.Bn. spots will present some unusual feature. In 1914, for instance, some dark oval spots on the N. component developed bright nuclei, the effect produced being that of a number of links on a chain. In 1919–20 also a very large, dark, elliptical spot was frequently seen to have a minute white spot at its centre.

Throughout the apparition of 1928–29 there persisted a conspicuous N.E.Bn. configuration, consisting of two intensely dark, elongated spots separated by a round white one, such as has been mentioned above. The distance from the centre of the white spot to that of the preceding dark one was about 9° and to that of the following one 11° of longitude. Some distance following these was a similar dark elongated spot, followed again, after a considerable gap, by another white one. The last was seen to be gaining rapidly on the dark spot preceding it; and this relative motion continued until the white spot was following the dark one by about 10°, when it was arrested, after which the distance between them remained approximately constant. Meanwhile yet another very dark, elongated spot had developed and was seen to be overhauling even the rapidly moving white one, which it approached to within about 12° before its velocity too was arrested and a replica of the first configuration became established; the two groups then presented a striking appearance, as they could be seen on the disk at the same time.

The mean rotation period of the N. Tropical Current, in which all these features move, for the forty-two apparitions between 1895–96 and 1947 inclusive for which it was determined, was $9^h 55^m 29^s$. The mean period of all the spots of a single apparition was longest in 1940–41, when its value from twenty-one spots was $9^h 55^m 39^s$, and shortest in 1938, when fourteen spots gave $9^h 55^m 13^s$. It will be seen, therefore, that it is far from being constant; moreover, during the majority of apparitions there have been one or two outstanding spots that have been remarkable for the rapidity with which they have moved in the direction of decreasing longitude, as for instance the white round spot and the dark elongated one just described. The record seems to be held by a small, very dark spot, whose mean rotation period from 1942 December 19 to 1943 May 6 was $9^h 55^m 3^s$ and from January 26 to May 6, after receiving an acceleration, $9^h 55^m 0^s$. Its supremacy is, however, closely contested by a dark spot of 1927–28 that exhibited for thirty-two days a period of $9^h 55^m 1^s$. At the other end of the scale, the N. Tropical spot with the longest observed period was a white one, recorded on five occasions only from 1929 August 25 to September 11, whose period of $9^h 56^m 4^s$ was equal to that of the slowest of the four dark spots of that apparition at the N. edge of the N.T.B. From the longitude chart it appears that this spot must have passed over (or under, or round) one of the dark N. Tropical spots during its short history and its identification after this event may not rest on a very sound basis; but three other white spots in the same region gave a mean period of $9^h 55^m 59^s$ and two of these may also have encountered and passed by a dark spot that had a much shorter period.

The apparitions just cited provide examples of a not uncommon feature of the N. Tropical Current, namely its division, longitudinally, into fast and slow moving sections. The following instances are taken from the *B.A.A. Memoirs* since 1900. For some time during the apparition of that year a portion of the N.E.B. became very vague and indistinct, especially along its northern edge, and at the same time the spots bordering on the N. Tropical Zone in this region, which covered about 90° of longitude, appeared to Phillips as though they were veiled by mist. As the normal drift of the spots brought them into this section, their rotation periods suddenly decreased by varying amounts, the more rapid motions persisting so long as they remained in the section, the boundaries of which were quite sharply defined. P. B. Molesworth found mean periods of $9^h 55^m 30^s$ and $9^h 55^m 21^s$ for what he respectively termed the normal and the abnormal drifts. In 1922 two white spots, whose original periods were only about $9^h 55^m 5^s$, had evidently been crossing such a region; for towards the end of May their periods lengthened suddenly by about 20^s, to become more nearly comparable with those of the rest of the N. Tropical spots. In 1927–28 there was a region, not more than about 90° in length, in which the mean rotation period was $9^h 55^m 44^s$, while spots in the remainder of the zone gave about $9^h 55^m 28^s$.

The apparition of 1929–30 was remarkable, in that there were two regions of long rotation period and two of short. On the date of opposition, 1929 December 3, the first definite short-period section ran from about System II longitude 350° to 40°, to be followed by a long-period section from 40° to 140°; then came another short-period section from 140° to 270°. In the remaining 80° more long-period spots were recorded; but this region seemed to be contested by one or two spots moving with the shorter period which, as already mentioned, may have succeeded in passing the others if the identifica-tions on the chart have been correctly made, though no actual observation of such an occurrence was obtained. In 1939–40 there were some interesting bright spots in the N. Tropical Zone and a rather inactive region covering about 60° of longitude. One of these bright spots, when first seen, was at the following end of the inactive region but was endowed with a rotation period, shorter by some 14^s than the mean of the N. Tropical spots, which carried it in solitary majesty through the deserted area. As it approached the preceding frontier, there were indications that it had begun to move more slowly; but as the end of the apparition was at hand, its subsequent behaviour is uncertain.

We return for a moment to the interesting behaviour of the record-breaking little dark spot of 1942 December 19 to 1943 May 6. When first seen and until the end of February, this spot lay only a little to the north of the middle of the N.E.B.; somewhere about the latter time, however, it seems to have moved northwards, for thereafter it was always recorded as lying at the N. edge of the belt. Now a change in the latitude of an atmospheric configuration that brings it nearer to the pole of a rotating planet might be expected to be accompanied by a decrease of its rotation period; the interesting point is that, although such an acceleration was observed, it took place, not at the beginning of March, but towards the end of January.

So many of the N. Tropical spots have been conspicuous objects that the question of their duration as individuals is of great interest. In numerous instances there has been little change of appearance throughout the whole of one apparition; but the comparatively long intervals when Jupiter is too close to the Sun for observation make identifications from one apparition to the next extremely difficult to claim with confidence, particularly when it is remembered that the rotation periods of these objects are subject to sudden and quite appreciable changes. The fact that the extrapolated motion in longitude of a white spot, say, of one apparition brings it close to the position where a white spot is observed at the beginning of the next is not, therefore, a very reliable criterion for the identity of the two objects. The author has been concerned with plotting longitudes and deriving rotation periods for the spots of twenty or more apparitions; yet in one instance only, that of a white spot that survived from 1941 August 18 to 1943 May 27, has he felt complete confidence in the identity of one of the N. Tropical spots with a similar feature of the preceding apparition; he has no doubt whatever that some of these objects have been possessed of considerable longevity but feels that great caution should be exercised before such a claim is accepted for any specified individual spot. In the case just cited the interval without an observation was only just over four months, from April 30 to September 9, yet although the preceding and following intervals were of similar duration, no trace of the spot is to be found in the records of either 1940–41 or 1943–44. It will be seen from the table of rotation periods on p. 100 that between 1940–41 and 1941–42 the rate of the N. Tropical Current underwent a somewhat sudden change.

Phillips, whose judgment merits the greatest respect, referred in 1900 to a white N. Tropical spot as being his Spot No. 24 of the previous apparition and in 1917–18 stated that one of the dark spots was 'probably identical' with one of the 1916–17 objects. Truly amazing claims come, however, from Denning and Williams with regard to two bright spots, one of which is stated to have persisted from 1887 till 1891 and the other from 1885 till 1891.* It is possible to examine the evidence upon which was based the identification of these two objects over the interval of nearly nine months, that elapsed between the last observations of 1887 and the first of 1888, by reference to Vols. I and II of A. S. Williams' *Zenographical Fragments* (now out of print). The spots are designated 'L' and 'C' respectively and those who have access to the originals should note that in 1887 they are classed as 'N. Temperate' objects, the modern nomenclature not having been adopted by Williams when the first volume was published. In the case of the spot 'L' its longitude, when first observed in 1888, was 15° away from that predicted by extra-polation from 1887 and Williams stated that it was mainly its great brilliance upon which the identification depended. The spot 'C' was first seen in 1888 within 8° of the extrapolated position; but its rotation period had shortened by six seconds; so this very fair agreement is significant, only if the change in its rate of drift took place rather suddenly just before 1888 March 25. Again Williams cites as evidence the great brightness during both the apparitions; but these spots often rise to prominence or fade away fairly rapidly, and so, taking also into consideration that there were nine white spots recorded on the zone in 1887 and twelve in 1888, the reader may feel inclined to share the author's scepticism as to the validity of the identifications claimed.

It is convenient to study next the southern margin of the belt, in order that a clear picture may be obtained of what is happening

* *B.A.A. Memoirs*, Vol. I, Part V, 80, 1893.

at its two edges before we consider the interesting phenomena that are often presented by its central regions. The dark spots and projections and the light rifts that occur so frequently at the S. edge of the N.E.B., invariably move with the northern portion of the great Equatorial Current which has a mean rotation period that is some five minutes shorter than that of the N. Tropical Current. Thus, if two spots, one at the S. edge and one at the N. edge of the belt, appear to have the same longitude, then, after five revolutions of the planet, which are completed in just over two days, the southern spot will have advanced relatively through about $15\frac{1}{2}°$ in the direction of decreasing longitude and will arrive on the central meridian twenty-five minutes earlier than the northern one. The author well remembers the thrill he experienced in discovering this effect for himself soon after he began the serious study of Jupiter, when he was under the impression that System I longitudes were applicable only to spots that lay actually in the Equatorial Zone. If two such spots should be suitably situated at opposite edges of the belt, a 3-inch refractor will readily reveal their rapid relative motion.

The dark projections are generally features that catch the eye immediately. They take many forms, from tiny humps or short spikes to large elongated masses or streaks. The humps and spikes are often the points of departure of grey wisps or festoons, some of them most delicate and some quite conspicuous, that seem to issue from the S. edge of the belt and look as if they were dispersing like smoke in the Equatorial Zone. Frequently, however, they do not simply vanish but curve round (no apparent motion is implied by these attempts at simile) and return to the belt, almost certainly reaching it at a point where another projection appears. Sometimes a wisp will curve right over one projection and return to the second one following its point of departure. Quite often one of them will fork into both preceding and following directions; this may lead to the formation of a series of grey arches with light, or even bright, central regions, the whole presenting a most fascinating spectacle, and any such light oval area may contain a bright nucleus, which, however, is seldom to be found near its centre but more probably fairly close to the belt near one of its ends. If a projection has been by-passed by one of the curving wisps, the enclosed light area will not, of course, be elliptical but will assume a shape resembling that of a kidney bean.

The dark masses and streaks may be strikingly conspicuous and are sometimes surprisingly rectangular in outline. It is not uncommon for such an object to be completely separated from the belt by an exceedingly narrow light line, this appearance being more readily seen by some observers if the head and eyes are turned into the position from which the belts are apparently arranged vertically. When one of the dusky wisps is associated with an elongated projection, it nearly always, though not invariably, issues from one of its ends. Further reference to these wisps is deferred, as we have already trespassed somewhat extensively upon the territory of the Equatorial Zone to which the next chapter is allotted.

Some idea of the development of one of the N.E.Bs. projections may be obtained from the following sequence, which is taken from the author's observing book:

Date		Longitude System I
1941		°
Dec. 15	Point of departure from N.E.Bs. of grey wisp curving f. to Equatorial Band	140
17	White dent in N.E.Bs.	142
18	White dent in N.E.Bs.	141
22	Small projection N.E.Bs.	140
24	Conspicuous tall projection N.E.Bs.	141

There is little doubt that the projection grew enormously between December 22 and 24. The white dent, which seems to have been a very temporary affair, evidently moved right across the longitude of the projection; for it was first seen on December 15 at 144°, just following the grey wisp, while on December 22 and 24 it was recorded at 136°, after which it was not seen again.

Since the dark projections often form a long series, that may even encircle the planet, there must naturally be light 'bays' between them. These bays frequently take the form of, or lead into, white rifts that cut deeply into the S. edge of the belt, sometimes breaking right through the S. component, when the belt is double, and leading into the light space between the components. If the rift curves as it enters the light space, the curvature is nearly always, though not quite invariably, in the 'following' direction; it may even turn northwards again and break through the N. edge of the belt into the N. Tropical Zone, possibly emerging close to one of the white spots, and the aspect will then resemble that of a rather obtuse 'S' bend.

Often, even when the belt is single, light rifts are to be seen running right across it; they may be quite straight but are seldom orientated along a meridian, the preceding end being with few exceptions the southerly one. This preferential orientation suggests that a rift may be formed originally at the time when there is conjunction in longitude between two white spots or gaps, one at the N. and one at the S. edge of the belt, which are then drawn out into a streak as the differential motion of the two currents in which they move causes the southerly spot to run ahead in the direction of decreasing longitude. It may be thought that this suggestion should be easy enough to verify; but in fact it is one that calls for considerably more investigation. The difficulty lies in the rapidity with which the spots separate after conjunction, which, as has been stated, is at the rate of 15° in two days, this interval being the minimum after which it is usually practicable to re-observe any particular marking near the central meridian. Unfortunately this rate is just a little too slow for a significant change to be recorded during a single presentation of the feature. Excluding the possibility of daylight observation, which might be successful if the rift were a bright one and if the light of the sky were reduced by means of a suitable filter, or by polarisation if Jupiter were near quadrature, the best chance of acquiring the necessary data seems to await an observer who is situated in fairly high terrestrial latitude when the planet is in high declination and who is lucky enough to catch a pair of spots in conjunction at about nineteen hours, local time, one evening; if the sky remains clear, he will have the opportunity of re-observing the configuration at 5 hours on the following morning, when, after just one revolution of the planet, the difference of longitude will have become 3°; after this he will have to wait until about 21 hours on the second evening following, before he can examine the region again for further evidence of development.

The central regions of the N.E.B. present at times a scene of great activity, which generally takes the form of an outbreak of white spots or rift-like streaks; some of these may be quite brilliant, yet in a typical region every improvement in the seeing conditions will reveal new delicacies of structure, the beauty of which is hardly to be rivalled elsewhere on the planet. The rifts may be quite short and either straight or slightly curved—in 1927–28 one of them, from its shape, acquired the nickname of 'banana'—and under the best definition they nearly always show bright nuclei; it may even be that most of them really consist of chains of little white spots, almost in contact, which can be revealed, even by large instruments, only when the seeing is superb. Lying, as they do, in a latitude that falls between those dominated by the N. Tropical and the N. Equatorial currents, it is not surprising that the rotation periods of these objects are also intermediate; but these differ widely among the individual spots, which are generally rather short-lived, and since changes of form may occur very rapidly, accuracy in the determination of their periods is not always easy to attain. In 1927–28 the author * found a tendency for the period to lengthen after a spot had been in existence for a few days. Plate IV shows a series of drawings by three observers of the same region from 1927 September 26 to October 22 and illustrates well the rapidity with which changes were taking place. The arrow-head marks approximately the same position, relative to the active area; its advance on the white N. Tropical spot between October 10 and October 22 should be noted.

Two remarkable white mid-N.E.B. spots call for individual mention. The first, which was large enough to fill the whole width of the space between the components of the belt, was seen by Hargreaves on both 1941 December 27 and 28 to have a small and intensely dark spot at its centre (slightly eccentric on December 28)— see Plate V, Fig. 1. The second was smaller but conspicuous and, when first recorded on 1946 May 21, was gaining rapidly on a typical dark N.E.Bn. streak which, owing to the great width of the belt, was situated well within its north edge in the latitude occupied by the northern portion of the white spot. The two were in conjunction on May 28, when the author was fortunate enough to obtain a few moments of seeing that was good enough to reveal two little dark spots, one on either side of the northern half of the white one; these were the two ends of the dark streak, the middle of which was obscured. The white spot was not recorded again; but on May 31 Hargreaves observed the dark streak, which, by inference, was restored to its previous condition. The region was presented again on June 2 under very bad definition; the impression obtained was that a light area, that was following the streak on May 28, had spread over and partially obscured it. It was not recorded again.

The records of the colour estimates that have been made of the N.E.B., whether of the belt as a whole or of its edges and central regions separately, are such as to leave the investigator in a state of bewilderment. A. S. Williams' search for periodicity in the colora-tion of the two Equatorial Belts has already been discussed and the modern records reveal little to add to it. While it can be stated that the redness of the N.E.B. has attracted far more attention in some apparitions than in others, the estimates are very far from being consistent among themselves, possibly because they were dependent upon the longitudes presented at the times of observation. It must be remembered too that the Earth's atmosphere is a colour filter of

* B.A.A.J., Vol. 38, No. 2.

which the selectivity may vary from hour to hour, quite apart from the small dispersive effect. One of the most impressive features of the belt's colour scheme is the contrast that is so often recorded between the redness of the northern edge or component, including the dark spots or streaks associated with it which have been described as crimson, as ruby red and perhaps rather aptly as having the colour of a clot of dried blood, and the neutral or even bluish tints usually ascribed to the markings near its southern boundary. An observer whose acquaintance with the planet did not date back more than forty years would probably subscribe to the statement that, whenever there is any red at all in the belt, the N. edge is affected by it; but a surprise would be in store for him, were he to consult the records of 1906–07 and 1907–08. In 1906–07 the belt was characterised by a number of dark reddish streaks along its S. edge, while observers agreed that the N. component, which had been decidedly reddish at the beginning of the apparition, developed later a cool colour which ultimately became bluish-grey or steel blue. In the Thirteenth Report of the B.A.A. Jupiter Section, dealing with the apparition of 1907–08, Phillips, after referring to a progressive decrease in the intensity of the coloration, wrote: 'Thus, the N. component was undoubtedly blue throughout; the middle regions were orange till quite the later part of the apparition, and the S. component, which at first was brick red, ultimately faded to a dull reddish brown.'

Even after making the maximum allowance for the manifold effects that may be attributable to the Earth's atmosphere, against which all observers who make colour estimates must be on their guard, it is impossible to doubt that many of the records of the colours associated with the N.E.B. have been based upon reality. As evidence of this we may adduce the two apparitions just cited, when the normal tints of the two components were interchanged; also the fact that, when two projections at the S. edge of the belt have been simultaneously under observation, one of them has sometimes been described as strikingly blue and the other as neutral grey; on one occasion Hargreaves recorded a projection that was blue-green on one side and red-brown on the other. Reliable colour estimates, made with a reflector when Jupiter was near the zenith, would be valuable; unfortunately the original notes of P. B. Molesworth, who observed from Ceylon and who must have often been favoured with just these conditions, are not now available.

The following Tables give details of the rotation periods that have been derived for two of the currents associated with the N. Equatorial

Belt; the periods found for objects at its southern edge are tabulated at the end of the chapter dealing with the Equatorial Zone, since the drift they represent forms a part of what has been named the Great Equatorial Current.

Rotation Periods of Spots and Markings in N. Tropical Zone and N. part of N. Equatorial Belt (N. Tropical Current)

Apparition	Rotation Period	No. of Spots
1787	9^h 55^m 34^s	
1835	24	
1866	18	
1881	35	
1887	36	17
1888	41	18
1890	34	5
1891	27	
1894–95	35	9
1895–96	31	2
1897–98	25	5
1898–99	31	21
1900	28	26
1901	30	28
1902	29	13
1903	31	13
1904–05	32	20
1905–06	26	21
1907–08	22	27
1908–09	26	11
1909–10	30(?)	17
1911	16	3
1914	32	18
1915	28	25
1916–17	30	21
1917–18	28	9
1918–19	28	10
1919–20	31	9
1920–21	28	2
1922	21	7
1923	24	4
1926	32	8
1927–28	34	19
1928–29	30	12
1929–30	38	24
1930–31	29	11
1931–32	36	22
1932–33	28	16
1934	28	7
1935	36	5
1936	30	6
1937	30	9
1938	13	14

Rotation Periods of N. Tropical Spots and Markings (*continued*)

Apparition	Rotation Period	No. of Spots
1939–40	27	15
1940–41	39	21
1941–42	27	24
1942–43	30	28
1943–44	28	6
1944–45	33	14
1946	28	10
1947	19	3
1948	9 55 26	4
	Mean $9^h\ 55^m\ 29^s$	

The figure given for 1909–10 is based on the observations of E. Hawks and Phillips, who were in close agreement; S. Bolton, however, derived from his own observations a period that was about 11 seconds shorter.

Rotation Periods of Spots in Middle of N. Equatorial Belt

Apparition	Rotation Period	No. of Spots
1898–99	$9^h\ 55^m\ 35^s$	10
1900	55 29	10
1927–28	53 29	12
1931–32	54 32	4
1932–33	54 44	2
1941–42	54 59	4
1944–45	53 50(?)	
1946	53 50±	
1948	9 53 40	1

It is worthy of note that the shorter period associated with this region does not appear to have been noticed before 1927–28.

CHAPTER 11

THE EQUATORIAL ZONE

Within this great zone lies, on the average, about one-eighth of the entire surface of the planet. As its name implies, it bestrides the equator, about which, however, it is seldom situated quite symmetrically. The lack of symmetry does not appear to have any correlation with the Jovian seasons and the latitude measurements show that it disappears in the long run, the mean values from 1908–09 to 1947 being $+7°2$ for the northern boundary and $-7°1$ for the southern. During this period the latitude of the northern edge varied between $+4°5$ in 1937 and $+9°0$ in 1944–45, that of the southern between $-4°4$ in 1919–20 and $-9°0$ in 1909–10; its greatest breadth was $17°6$ in 1914 and its least $11°0$ in 1924.

Threading its way, nearly along the middle of the zone, telescopes of moderate aperture will often reveal the thin, faint, dusky line, known as the Equatorial Band. Sometimes this will run right across the visible disk; more often it is fragmentary and it seldom encircles the planet. On the two occasions for which latitude measurements are available the Equatorial Band lay within 1° of the true equator; if, however, its placing upon drawings is to be relied upon, this is by no means always the case, though its departure from the equator is naturally never pronounced.

During the three consecutive apparitions of 1937, 1938 and 1939–40 the Equatorial Band was unusually prominent. In 1937 it was so wide in some longitudes as to be almost like a belt and on one occasion it was described as resembling the N. component of a belt, of which the S. component was the S.E.Bn. A note, made by the author on 1938 June 29, reads: '. . . there is a heavy dark line of about the thickness and intensity of the N. component of the S.E.B., but in the position of the E. Band. There is no N. component of the S.E.B. here (121°–132° System I) but it starts again in more or less normal latitude at 135°.' In 1939–40 the Band was often seen to be very wide and dark, sometimes wider and stronger than the N. component of the S.E.B., and on at least two occasions it was recorded as being double.

Frequently one or more of the grey wisps, originating in a projection at the S. edge of the N.E.B., will curve over and, instead of returning to the N.E.B. in the form of an arch, will merge into or initiate a section of the Band. Wisps and loops from the N. edge of the S.E.B. are less frequent; but they do occur and may behave in a similar manner, or they may even run right across the zone to join the N.E.B. at some salient point.

Having returned to the grey wisps, we may ponder for a moment on the nature of these delicate and rather beautiful filaments. When they take the form of loops or arches, enclosing light areas, some observers tend to see the light areas as the dominant features; and this raises the question of whether the wisps can be contrast effects, induced at the edges of brighter regions. To most observers, the author among them, they certainly look objective enough; but those who study illustrations of them should remember that planetary drawings are made for convenience with pencil on white paper and that part of the technique of portraying a lighter area on a white background is to create an optical illusion by over-emphasising the contrast at its edge. Possibly the eye may allow the converse of this trick to be played upon it, when a light object is viewed against a background that is only slightly less luminous. It seems not unlikely that, if the usual practice were for drawings of Jupiter to be made with chalk upon dark paper, the draughtsman might acquire a different outlook as to which were really the objective features and that his mental picture, when actually observing with the telescope, might undergo some revision. It is improbable that this point will be readily decided; but the alternative seems to be one that is worth bearing in mind, however strong the conviction may be that the wisps are objective grey streaks.

There is often a fair amount of other detail, in the form of light spots and dusky patches, to be made out on the Equatorial Zone; but except at times of special activity, such as occurred in 1938 and 1939–40, it may require first-class definition to reveal the rather delicate mottlings that may be present. The zone as a whole is subject to pronounced changes of intensity and will vary all the way from being the brightest region of the planet to assuming such a sombre hue that, at first glance, the impression may be received that the N.E.B., the E.Z. and the S.E.B. comprise one vast belt. The summary that follows indicates the apparitions since 1900, during which either the intensity or the activity of the zone seems to have been worthy of special comment.

102

1903	N. part of zone brighter than S. part.
1904	S. part of zone brighter than N. part.
1905–06	Darker than either of the Tropical zones. Light cloud-like patches in S. half.
1908–09	Very bright, especially at the beginning of the apparition.
1912	Bright.
1913	The S. part gradually darkened, leaving a series of regular bright ovals in the N. half of the zone, some of which were brilliant in the extreme.
1914	S. part very dark at first, becoming lighter. Light ovals in N. half.
1916–17	Rather bright.
1917–18	Rather bright.
1919–20	Strongly shaded.
1920–21	Decidedly darker than either of the Tropical zones.
1923	White.
1924	White.
1925	Dull; but some brighter areas, especially near N.E.Bs.
1926	Lightly shaded.
1927–28	Shaded; but with narrow, bright S. edge. Some light northern areas.
1931–32	Very bright.
1932–33	Very bright.
1936	Somewhat shaded, particularly the S. part, but much broken up into bright oval areas.
1937	No discontinuity at times between the dusky S. part of the zone and the faint N. comp. S.E.B. A large number of white spots in the N. part, sometimes so congested as to make the whole region look white.
1938	A scene of great activity. S. part dusky, with several large dark masses. These, together with other features in the same latitude, had originally rotation periods of about $9^h 51^m 40^s$; but after a while a steady acceleration set in, which in little more than two months had reduced the periods to $9^h 50^m 30^s$, this figure remaining nearly constant for the remainder of the apparition. In the N. part were many long bright oval areas, which in good seeing could generally be resolved into complex formations of two, and sometimes three, white spots, arranged nearly longitudinally, of which the preceding member was frequently the smallest, brightest and most northerly.

1939–40   S. part darker than in 1938 but less disturbed; again sometimes seemed to merge with S.E.Bn. N. part intensely active; chains of white spots again in evidence.

1940–41   Bright; southern duskiness gone.

1941–42   Bright at first; then fading somewhat.

1942–43   White spots numerous in N. half; a few, more diffuse, filling almost the whole width of zone.

From a study of the very numerous records of the colour of the Equatorial Zone the following points seem to emerge:

(1) When the zone is bright, its colour is nearly always white.

(2) When it is dusky, the colour may exhibit one of two alternative tendencies. Records of the less frequent of these refer to the zone as having a dull metallic or leaden hue; but far more commonly expressions such as ochre, brownish-yellow, pinkish and even rosy are to be found. The term which seems best to summarise the latter group and which has frequently been used by observers is 'tawny'.

On the title page of Agnes M. Clerke's *History of Astronomy*, published in 1893, appear reproductions of two early photographs, one of Jupiter in 1879, the other of Saturn in 1885. The picture of Jupiter shows the Great Red Spot, as an intensely dark elongated object, and a single conspicuous dark belt, apparently on the equator. The aspect of the latter will be completely baffling to the modern student of Jupiter, who may wonder how the transformation from a dark 'Equatorial Belt' with a light region on either side of it to a light Equatorial Zone bounded by the dark N.E.B. and S.E.B., with which he is familiar, can have come about in so comparatively short a time; until he realises that the wide 'belt' in the illustration includes the S. component of the N.E.B. and the N. component of the S.E.B. and that the Equatorial Zone was so deficient in blue and violet light that it registered photographically as an equally dark object. There seems little doubt that visual observers in 1879 saw the zone with a decidedly orange or pinkish tinge; some tinted drawings in the author's possession that were executed in 1880 by Henry Corder certainly present the normal appearance but the E.Z. is heavily shaded with pencil and washed over with pale ochre.

One remarkable instance of local coloration must be mentioned. On 1928 December 8 Hargreaves and Phillips, observing together and employing an 18-inch reflector, noticed between the centre and N. edge of the Equatorial Zone quite a dusky area of the most intense blue; the region immediately to the south of it was red-brown. Phillips' note reads: 'A very remarkable large blue shading or area was seen on the E.Z. . . . Hargreaves described it as Prussian Blue. It has a distinct tinge of green in it. At the same time the S.E.B. was seen to be very definitely red, and the whole effect was very striking indeed.' The same area was conspicuous again on the following night, when Hargreaves wrote of it: 'This now looks brilliantly blue.'

The tables that follow give the rotation periods that are available for the three branches of the Great Equatorial Current for a large number of apparitions. The northern branch includes the features at the S. edge of the N.E.B. that have already been described, the central branch relates to spots and streaks in the middle of the Equatorial Zone and the southern branch, while embracing spots in the E.Zs., includes in anticipation most of the objects associated with the N. component of the S.E.B., which forms part of the subject matter of the next chapter. It will be noted that for about a decade, beginning with the apparition of 1898–99, the number of spots employed in deducing the periods was often large, which might imply that specially high weights should be given to the figures relating to those apparitions; actually, however, the reverse is probably the case, for most of the periods were due to a single observer, who recorded a great profusion of detail both conspicuous and delicate, with the result that the overcrowding of his charts with too many entries that were insignificant may easily have led, as has been suspected, to some errors in identification.

It will be seen that, although the figures vary considerably from apparition to apparition, there is no tendency to periodicity; nor are there any pronounced trends, except in the case of the southern branch, of which the period showed a fairly steady increase from 1879 to 1888, as was first pointed out by A. S. Williams in 1896. From this point onwards there seems to have been no further tendency for the period to lengthen, until in 1917–18 we come suddenly upon the figures 9^h 50^m 41^s for one spot and 9^h 51^m 8^s for the mean of three others. Subsequent records seem to suggest that, since the beginning of this apparition, the motion of the S. Equatorial and S.E.Bn. spots has been under the control of two separate influences, one of them tending to preserve the normal rate of drift that had fluctuated but little since 1888, the other to produce rotation periods considerably in excess of 9^h 51^m; the possibility must not be overlooked, however, that long periods in some of the earlier apparitions may have escaped detection. The two currents, if such there be, show no clear-cut division, except when they are in evidence simultaneously, and the author, in designating them 'A' and 'B', has chosen the somewhat arbitrary figure of $9^h 51^m 0^s$ as the shortest period for which he has assigned a spot to Current B. In 1928–29, during the great revival of the S. Equatorial Belt, most of the S. Equatorial spots had very long periods; and for a week or so two spots that lay almost in the middle of the Equatorial Zone were caught up in this slow rate of drift.

In 1911, for some reason, all the spots in the Great Equatorial Current were classified as simply 'Equatorial'; in preference to omitting them altogether, their mean period has been inserted in brackets under the central branch of the current and no entry made under the others.

THE GREAT EQUATORIAL CURRENT
Northern Branch

Rotation Periods of Spots at S. edge of N. Equatorial Belt and in N. part of Equatorial Zone

Apparition	Rotation Period	No. of Spots
1882	$9^h 50^m 40^s$	
1884–85	9	
1888	24	27
1891	32	
1896–97	36	2
1897–98	25	3
1898–99	26	27
1900	27	42
1901	25	44
1902	uncertain	
1903	uncertain	
1904–05	29	24
1905–06	40±	34
1907–08	34	27
1908–09	27	16
1909–10	32	25
1913	12	28
1914	17	22
1915	15	23
1916–17	14	14
1917–18	19	30
1918–19	19	22
1919–20	17	6
1920–21	34	10
1922	26	12
1923	10	10
1924	13	8
1925	21	8
1926	23	28
1927–28	24	32

THE GREAT EQUATORIAL CURRENT
Northern Branch (*continued*)

*Rotation Periods of Spots at S. edge of N. Equatorial Belt
and in N. part of Equatorial Zone*

Apparition	Rotation Period	No. of Spots
1928–29	24	53
1929–30	17	29
1930–31	26	29
1931–32	26	35
1932–33	26	27
1934	22	20
1935	29	25
1936	28	18
1937	26	19
1938	21	33
1939–40	22	37
1940–41	20	36
1941–42	22	28
1942–43	21	28
1943–44	20	16
1944–45	22	31
1946	23	13
1947	24	7
1948	9 50 24	9
	Mean $9^h\ 50^m\ 24^s$	

THE GREAT EQUATORIAL CURRENT
Central Branch

Apparition	Rotation Period	No. of Spots
1898–99	$9^h\ 50^m\ 29^s$	8
1900	26	13
1901	28	9
1903	25	18
1904–05	29	47
1909–10	28	11
1911	(32)	7
1927–28	28	6
1928–29	⎰ 50 28	7 ⎱
	⎱ 52 15*	2 ⎰
1932–33	50 32	1
1934	50 30	3
1938	49 46	1
1939–40	50 24	6
1942–43	9 50 20	2
	Mean $9^h\ 50^m\ 25^s$	

* Not included in mean.

THE GREAT EQUATORIAL CURRENT
Southern Branch
Rotation Periods of Spots at N. edge of S. Equatorial Belt and in S. part of Equatorial Zone

Apparition	A		B	
	Rotation Period	No. of Spots	Rotation Period	No. of Spots
1879–80	9ʰ 50ᵐ 1ˢ			
1880–81	6			
1881–82	10			
1882–83	10			
1883–84	13			
1884–85	13			
1885–86	23			
1886–87	22	21		
1888	32			
1891	26			
1897–98	24	19		
1898–99	24	42		
1900	23	46		
1901	29	28		
1902	17	24		
1903	23	42		
1904–05	29	48		
1905–06	28	60		
1907–08	28	42		
1908–09	26	24		
1909–10	26	37		
1914	18	1		
1915	29	1		
1916–17	27	2		
1917–18	41	1	9ʰ 51ᵐ 8ˢ	3
1918–19	47	2		
1919–20			0	3
1920–21	36	4	45	1
1922			15	4
1923	48	4		
1925	27	1		
1927–28	19	1		
1928–29			17	38
1929–30	52	13		
1930–31			6	6
1931–32			4	8
1934	29	8		
1938	30*		40†	
1939–40	50 28	5		
1942–43	⎰ 49 56	4 ⎱	9	1
	⎱ 50 37	1 ⎰		
1943–44	31	7	29(?)	1
1944–45	59	4	9 51 48	1
1947	30	1		
1948	9 50 29	2		
Means	9ʰ 50ᵐ 26ˢ		9ʰ 51ᵐ 21ˢ	

* After acceleration ⎱ These values are
† Before acceleration ⎰ approximate.

The mean values are not very significant, owing to the arbitrary nature of the subdivision.

CHAPTER 12

THE SOUTH EQUATORIAL BELT AND THE SOUTH TROPICAL ZONE

The South Equatorial Belt

Examination of the records leads to the conclusion that from 1892–93 to 1907–08 the S. Equatorial Belt was the most conspicuous belt on the planet during no fewer than eleven of the fourteen apparitions represented. It is indeed remarkable, therefore, that since 1908 it has been unquestionably the most prominent belt only in 1912 and in 1925; it has also been in general far less active than the N.E.B., as a comparison of the numbers of spots that have been available for the determination of rotation periods will confirm. Its general behaviour over the last forty years may be summarised by saying that there have been long periods of quiescence, punctuated by outbursts of intense activity such as have never been rivalled by the N.E.B. These outbursts followed periods when the S. component of the belt had either disappeared or faded to insignificance and took place in 1919–20, 1928–29 and 1942–43. It seems probable that there was also a minor upheaval while Jupiter was near the Sun between the apparitions of 1937 and 1938 and there was undoubtedly another major outburst in 1949*; but on the last occasion Jupiter lay in high southern declination and details of its progress seem to have been studied systematically by one observer only, R. A. McIntosh of Auckland, New Zealand.

These revivals of the S.E.B. have been so remarkable and the phenomena accompanying them so striking and unexpected, that a separate chapter is devoted to them later. Other impressive features, such as the prominent bay or indentation made in the southern portion of the belt by the Great Red Spot, which is known as the Red Spot Hollow, and the rapidly retrograding spots and projections at the S. edge of the belt, which are associated with the northern branch of the so-called 'Circulating Current', will also be

* More recently 1952–53 has provided further activity.

discussed in the chapters allotted to these interesting phenomena; this section, therefore, will be confined to a short review of those features of the belt whose behaviour bears no manifest relation to influences originating in the neighbouring regions of the planet.

The S. Equatorial Belt nearly always consists of two separate components, the space between them being often so wide, bright and featureless, that it might almost lay claim to be regarded as one of the recognised 'zones'. Sometimes, however, the space is not present in all longitudes, when portions of the belt will appear as a single broad band; but it is rare for the separation of the components to be invisible all round the planet. The width of the belt, as measured from the N. edge of the N. component to the S. edge of the S. component, is on the average somewhat greater than that of the N.E.B. and does not fluctuate in such a marked degree, though there have been occasions, as in 1937, when the S. component has completely disappeared for a time.

Most of the spots associated with the N. component of the S.E.B. move with one or other of the subdivisions of the southern branch of the Great Equatorial Current, of which the rotation periods have already been tabulated; a few that have projected southwards into the space between the components, have had longer periods. Many of the former have projected northwards from the N. edge of the component and some have been preceded or followed by light rifts, breaking through into the central space, not unlike their counterparts at the S. edge of the N.E.B. Some of the projections have been accompanied by grey south equatorial wisps; but since 1911 projections and wisps from the S.E.Bn. have been far less numerous than from the N.E.Bs. and in many apparitions have not been recorded at all; it is rather astonishing, therefore, to look back at the records and drawings of the first decade of the twentieth century and to note that during this period these features were displayed in far greater profusion by the S.E.B. than by the N.E.B.

Except during the great revivals, the light space between the components is normally tranquil; white spots and transverse grey streaks do appear, however, and may resemble superficially some of the mid-N.E.B. markings, though small brilliant nuclei are not characteristic of the white spots between the components of the S.E.B., nor in general are rotation periods found here that are markedly shorter than that of System II. The following table gives the mean periods obtained for this latitude since 1907–08 and it will be seen that the shortest time of rotation, that of 1930–31, was only a few seconds less than 9^h 55^m.

Rotation Periods of Spots in Middle of S. Equatorial Belt

Apparition	Rotation Period	No. of Spots
1907–08	9^h 55^m 34^s	3
1911	35	1
1922	21	1
1929–30	55 38	2
1930–31	54 52	5
1931–32	55 33	2
1934	9 55 33	4

Mean   9^h 55^m 27^s

During no other apparition were any spots prominent or permanent enough to furnish the requisite data.

In view of the results just tabulated, the following determinations, which are due entirely to Molesworth and relate to white spots in the same region, are somewhat disconcerting:

Apparition	Rotation Period	No. of Spots
1898–99	9^h 54^m 45^s	5
1900	51 37	13
1901	51 32	20
1903	9 51 27	16

The fact that one observer, from his own records alone, was able to derive rotation periods for so many spots, when others failed to contribute a single period, requires some explanation. For the first three of these apparitions this might lie in the fact that Jupiter lay well to the south of the equator, so that an observer in Ceylon would have had great advantages over those in the British Isles; but in 1903 the planet was nearly on the equator and not too badly placed for northern observers. Now Molesworth was a most assiduous and indefatigable worker, who recorded every detail he could see, whether conspicuous or elusive; his tally of transit observations for a single apparition sometimes exceeded 6000! It seems, moreover, that he plotted nearly all the derived longitudes, with the result that his charts were overcrowded. There is an optimum number and distribution of the points on such charts, if correct identifications are to be made in the absence of descriptive criteria for the spots in question; if too few points have been plotted, their identification is obviously attended by considerable risk; if too many, the manner in which they are connected may become quite arbitrary and erroneous periods may be all too easily deduced. Since most of Molesworth's spots were faint enough to have escaped detection by other observers, it is doubtful whether he was able to distinguish them individually by their aspects alone. The author is not the first to have suggested that the rather embarrassing discrepancies between some of Molesworth's published rotation periods and those deduced by his contemporaries were due to errors in identification; nevertheless, it is not without a feeling almost akin to impiety that he expresses a waning faith in results that were given by an observer, who made such a monumental contribution to the study of the Great Planet, and submits that the periods he found for the mid-S.E.B. spots of 1900, 1901 and 1903 may have been erroneous. It will be noted that the figure for 1898–99, when only five spots were employed, would be by no means out of place in the previous table.

In favour of Molesworth's results it must be admitted that the non-recurrence of such periods does not necessarily provide an adequate reason for doubting them—was there not an interval of thirty-seven years between the outbreak of rapidly moving N.T.Bs. spots of 1891 and their subsequent re-detection in 1929?—also that it is unscientific to reject observations simply because one does not like the look of them. Reference to the table of rotation periods for the S. branch of the Great Equatorial Current will show that, had the objects been on the N. component of the belt instead of in the middle, their periods would have been quite representative of the spots in Group B; moreover, if the spots had been associated with one of the great revivals of the belt, their motions would not have been unique. But they were not on the N. component; nor is there any question of there having been a revival, though the belt was strong and fairly active. Whatever the true facts, it is to be regretted that the figures representing the rotation periods of these interesting spots should be based upon the unconfirmed work of a single observer.

The S. component of the S. Equatorial Belt has been the scene of considerable activity during the great revivals of the belt, when it has provided the spots endowed with the longest rotation periods ever observed on the planet. With it are also associated the long period spots that define the northern branch of the 'Circulating Current'. In this section we are reviewing only its aspect during normal apparitions, when it is generally in rather a tranquil state. Its prominence has varied greatly during the last thirty-five years, far more than has that of the N. component; for when fading of the belt has set in, prior to one of the great revivals, the S. component has been reduced either to a thin faint line, to fragments or even to complete invisibility. The first modern instance of such a fading took place towards the end of the apparition of 1918–19; but the S. component was unquestionably faint during the years from 1878 to 1881, when the Great Red Spot was creating so much sensation. By the end of 1882 it was quite strong again.

Dark spots here normally take the form of condensations in the component or rather small projections from its S. edge; rifts or interruptions in the component provide most of the light ones. It is of interest and probably of the highest significance that there have never been recorded at the S. edge of the S.E.B. any features in the least analogous to the conspicuous dark elongated spots and large round white ones that are so characteristic of the N. edge of the N.E.B.; indeed, as we study the S. Tropical and S. Temperate regions of Jupiter, the lack of symmetry between the two hemispheres is very forcibly demonstrated.

In normal times the rotation periods of the S.E.Bs. spots do not differ markedly from that of System II, as the following table, from which have been excluded the periods of all objects associated with special phenomena, will show. It will be seen that the first entry is for 1911. For each of the five consecutive apparitions prior to this the mean of the published periods lies between $9^h 55^m 20^s$ and $9^h 55^m 25^s$; but the exact situation of some of the objects employed in their derivation is not clear, nor is their relation to the S. Tropical Disturbance whose period was similar. It is somewhat doubtful, therefore, whether a sudden change in the rotation period of the S.E.Bs. markings took place in 1911 or whether this is only apparent and due to the selection of different types of spot for discussion.

Rotation Periods of Normal Spots on S. Component of S. Equatorial Belt

Apparition	Rotation Period	No. of Spots
1911	$9^h 55^m 41^s$	1
1913	48	2
1915	35	1
1916–17	36	1
1917–18	32	1
1918–19	46	2
1926	39	2
1929–30	33	1
1931–32	36	3
1932–33	41	6
1935	28	2
1942–43	43	1
1943–44	9 55 49	3
Mean	$9^h 55^m 39^s$	

Very occasionally one or other of the components of the S.E.B. has been seen to be double in some longitudes, the S. component for instance in 1930–31 and in 1938 and the N. component in 1941–42. The indefinite N. edge of the N. component, where it practically merged with the dusky S. part of the Equatorial Zone in 1937, 1938 and 1939–40 has already been mentioned.

Colour estimates of the S. Equatorial Belt have been many and various and are almost impossible to summarise. Briefly, there have been apparitions when the belt has been described simply as grey, or even blue-grey; but far more often there has been some degree of redness present. The two components have by no means always been similarly tinted; in 1932–33, for instance, the N. component was considered to be grey and the S. component reddish-brown. A large number of estimates by individual observers has been published in the B.A.A. Jupiter Reports, which should be consulted by the student who requires more detailed information.

The South Tropical Zone

We have now reached the most remarkable region of the whole planet. That one zone, comprising only about 10° of latitude, should be the site of the Great Red Spot, the South Tropical Disturbance, the Circulating Current, the Oscillating Spots of 1940–41 and 1941–42 and the Dark South Tropical Streaks of 1941–42 and 1946 is little short of bewildering. Perhaps the most amazing fact of all is that the Red Spot and the S. Tropical Disturbance, during the forty years which were approximately the lifetime of the latter, not only occupied the same latitude but were many times in conjunction in longitude, the Disturbance, with its shorter rotation period, having streamed past the Red Spot again and again.

All the phenomena just mentioned will be discussed in detail later; so in this section, as in the last, we shall be considering only those features that relate to the zone in general.

It is perhaps hardly surprising that little is left for special comment. Until the S. Tropical Disturbance began to fade, the longitudes that it occupied were often heavily shaded; but in the undisturbed region the S. Tropical Zone has always been a conspicuous feature of the planet. During the majority of apparitions it has been bright, often the brightest part of the disk, and its colour has usually, but not invariably, been white. The whiteness has been striking at the times when the Red Spot has been especially conspicuous and illustrations make it clear that the zone was white and bright in 1880, when the Spot still maintained the prominence that in 1878 had made it famous.

The following table gives the rotation periods of various spots and markings observed on the S. Tropical Zone that do not seem to have been related to any special phenomenon. Some of them lay in the longitudes occupied at the time by the S. Tropical Disturbance, others in the undisturbed region; but the periods of spots in the middle reaches of the Disturbance do not often appear to be influenced by those found for its ends.

Rotation Periods of Spots and Markings
in S. Tropical Zone

Apparition	Rotation Period	No. of Spots
1925	$9^h\ 55^m\ 50^s$	1
1927–28	32	1
1928–29	22	4
1932–33	27	3
1934	50	4
1938	45	3
1948	9 55 29	3
	Mean $9^h\ 55^m\ 36^s$	

As in the case of the S. component of the S. Equatorial Belt the mean rotation periods of quite a large number of spots in the S. Tropical Zone were published for the five consecutive apparitions 1905–06 to 1909–10. The figures were practically identical with those given for the S.E.Bs. spots and for the S. Tropical Disturbance.

It may be noted that there is no single rate of drift in these latitudes that can be conveniently styled 'The S. Tropical Current'.

115

THE SOUTH TEMPERATE
BELT AND ZONE

The South Temperate Belt presents many features of outstanding interest. It is one of the permanent belts and is often conspicuous, having been many times recorded as the second most prominent belt on the disk; it has sometimes been the darkest of all, its inferior width having alone been responsible for its ranking as second to one of the Equatorial Belts. Frequently its intensity has varied very considerably with the longitude presented and over the last fifty years parts of it have been recorded as closely double during about one-half of the apparitions represented. Sometimes there has appeared a delicate grey line, which might perhaps be regarded as a thin component, lying a little to the north of the belt in the southern part of the S. Tropical Zone; and along this, when it has been present, have run the dusky spots that define the southern branch of the 'Circulating Current'. This grey line has been seen to issue in a south preceding direction from the preceding end of the S. Tropical Disturbance, whence it curved round to assume its course, parallel and close to the N. edge of the S.T.B.; from this appearance it acquired the name of 'The Smoke Stack'—see Plate V, Fig. 3 and Plate VII, Fig. 7.

Dark and light spots are frequent in this region and, together with a large number of preceding and following ends of darker, wider or double sections of the belt, have provided the material from which the rate of drift of the S. Temperate Current has been derived. In many modern apparitions p. and f. ends have predominated in the tables of objects for which individual periods have been determined.

In 1891 a very prominent dark section of the belt attracted a great deal of attention and was described by Williams as one of the most striking markings on the planet and also as being intensely red in colour. Under good definition this section of the belt was clearly seen to be double, a narrow bright stripe dividing it into two components of about equal width. It seems to have survived into the following apparition but to have retained neither its darkness nor its redness. Reference to Plate III, Figs. 3 and 4 will enable a comparison to be made with a very similar object that appeared in 1935; but it is to be remembered that neither of these markings is more than a conspicuous example of features that are common enough on the S.T.B. The dark section of 1935 was not recorded as double; but Jupiter was 16° south of the equator and seeing conditions may not have been adequate for northern observers. No mention of its having been red is found in the 1935 records and indeed there is little reference to redness in the S.T.B. at all during the present century; reddish-brown seems to have described its aspect during the four apparitions of 1936 and 1940–41 to 1942–43, while for the rest of the time it has been predominantly grey, and one may sometimes be tempted to wonder whether the appearance of the Great Red Spot in 1878 did not set a fashion for the somewhat indiscriminate application of the term 'red' to a number of Jovian markings, whose colour might have been more faithfully indicated by an expression such as 'warm brown'.

The South Temperate Current, under the influence of which objects associated with the S. Temperate Belt and Zone normally move, has been claimed as the steadiest of all the recognised drifts in Jupiter's atmosphere; but modern observations indicate that the S.S. Temperate Current is even more consistent, in rate of drift if not in range of latitude. At times the latter current, whose regular domain is the S.S. Temperate Belt and Zone, has extended its influence northwards and been responsible for the rotation periods of some of the spots even on the S.T.B. Until the end of the apparition of 1939–40 these incursions by the S.S. Temperate Current had been restricted as to their extent in longitude; but the apparitions after that, except for one S. Temperate marking in 1940–41 and one in 1941–42, were remarkable for the complete disappearance of the S. Temperate Current until at least the end of 1947* and the domination of all the S. Temperate features by the S.S. Temperate Current.

The table given below records the mean rotation periods that are available for spots and markings on the S. Temperate Belt and Zone, the figures in round brackets relating to objects that have been controlled by the S.S. Temperate Current. The latter have been excluded from the mean given at the foot of the table; and as some of the spots to which they relate have been re-grouped by the author,

* and more recently still; see Addendum to this chapter and Figure 3.

the remaining figures in the table will not always agree exactly with those that have been published previously for the S. Temperate Current.

Rotation Periods of Spots and Markings in the
S. Temperate Belt and Zone
(S. Temperate Current)

Apparition	Rotation Period	No. of Spots
1787	9^h 55^m 18^s	
1862	17	
1872–73	20	
1880–81	18	
1886–87	17	3
1887–88	(8)	1
1889	19	
1891	19	2
1892–93	21	1
1898–99	16	20
1900	17	46
1901	18	25
1902	19	7
1903	19	12
1904–05	20	11
1905–06	20	7
1906–07	22	20
1907–08	20	12
1908–09	20	17
1909–10	20	21
1911	{ 15	4 }
	{ (8)	2 }
1912	21	5
1913	23	1
1914	20	5
1915	21	4
1916–17	21	5
1917–18	19	3
1918–19	22	4
1919–20	{ 16	2 }
	{ (7)	2 }
1920–21	{ 22	2 }
	{ (6)	1 }
1922	15	2
1923	{ 16	3 }
	{ (8)	1 }
1925	23	3
1926	23	7
1927–28	23	13
1928–29	{ 22	7 }
	{ (10)	2 }
1929–30	23	8
1930–31	24	7
1931–32	20	10

Apparition	Rotation Period	No. of Spots
1932–33	24	13
1934	24	9
1935	17	3
1936	14(?)	
1937	25	1
1938	{ 19	8 }
	(10)	1 }
1939–40	22	4
1940–41	{ 14	1 }
	(8)	4 }
1941–42	{ 17	1 }
	(5)	9 }
1942–43	(6)	5
1943–44	(6)	10
1944–45	(5)	10
1946	(6)	5
1947	(8)	5
1948	(9 55 9)	9

Mean $9^h 55^m 20^s$

At about the same time as the S.S. Temperate Current began to take possession of the S. Temperate territory a remarkable series of wider or double portions of the S. Temperate Belt began to appear; and although similar features had been recorded before, it seems more than likely that the two phenomena were in some way connected. Where the belt was wider, the extension was in a southerly direction, so that the S. Temperate Zone became narrower in these longitudes, the limits of which, especially the preceding ones, were generally sharply defined by 'shoulders' on the S. edge of the belt— Plate V, Figs. 1 and 2. In some places the two separate components of the belt were fairly conspicuous, in others only the best seeing would reveal that a section of the belt, which had previously appeared simply as widened towards the south, was actually threaded down the middle by an exceedingly fine white line, a remarkable object if it is to be interpreted as a cloud formation; so it is not impossible that all the wider portions required only perfect resolution to display their duplex character.

The most noteworthy attribute of these features, however, was the length of time during which they remained conspicuous; apart from the Great Red Spot and Hollow, with which we include 'Hooke's Spot' of 1664 to 1713, and the South Tropical Disturbance they were the longest-lived objects for which the continued identity has been certainly established. In all there were three main wider or double portions of the belt, all of which could be traced from the beginning of the apparition of 1941–42 until the end of 1947, though

119

sometimes the preceding and sometimes the following end of one or other would fade, only to reappear again. Another remarkable transformation to which they were subject was a spreading out or diffusion, as it were, across the S. Temperate Zone; during some apparitions a section of the belt that had previously been double would not so appear, but the whole of the S.Temp.Z. in the longitudes it should have occupied would have grown dusky, with well-marked preceding and following ends that suggested miniature replicas of the p. and f. ends of the old S. Tropical Disturbance. Then the Zone would clear again and the double section of the belt be re-established. Such changes were actually seen to take place during the course of a few weeks.

The harbinger of the series appeared at the beginning of the apparition of 1939–40 as the preceding end of a thicker portion of the S.T.B. During the next two apparitions it took the form of the p. end of a dusky part of the S. Temperate Zone; but in 1942–43 it began to transform itself into a S. component of the S.T.B. and as such it remained until the end of 1947, when the zone to the south of it was beginning to grow dusky again. There seems to be no reasonable doubt that the same object persisted through all these years; but evidence of the progress of this and the other similar features is given by the longitude chart, reproduced as Figure 3, and in the following summary of the history of the whole series, in order that the reader may have the opportunity of forming his own estimate of the reliability of this claim to longevity.

Under the heading 'Character' in the summary the letter B denotes that the belt was wide or double, Z that the S. Temperate Zone was dusky, while B→Z or Z→B implies that a change took place during the apparition. The rotation periods given are those derived for the respective apparitions, and the periods between apparitions are reckoned over the interval that elapsed between the last observation of one apparition and the first observation of the next. A slight tendency may be discerned for the periods to be shorter between than during apparitions. This effect was noted by Phillips and confirmed by Stanley Williams many years ago in connection with the Red Spot; it is of the order of only a second or so and seems to be due to the influence of the planet's phase upon the visual estimation of transit times. The geometrical phase is allowed for in the ephemerides giving the longitudes of the central meridian, but not the fact that the terminator is more darkened than the limb; the latter effect must tend to cause the times of central meridian passages to be estimated too early before and too late after opposition, with

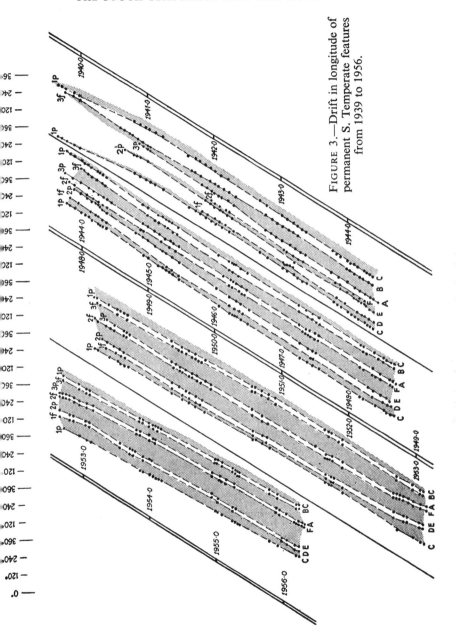

FIGURE 3.—Drift in longitude of permanent S. Temperate features from 1939 to 1956.

the result that the rotation periods deduced from them for a single apparition will come out a trifle too long.

Summary of the History of Long-lived S. Temperate Markings

Apparition	Character	Longitude at Opposition	Rotation Period	Rotation Period between Apparitions
		Preceding End of No. 1 (1p on Chart)		
1939–40	B	267°	9h 55m 14s	
1940–41	Z	14	10	9h 55m 14s
1941–42	Z	55	7	6
1942–43	Z→B	74	6	5
1943–44	B	80	5	3
1944–45	B	84	4	3
1946	B	106	8	6
1947	B and Z	136	9 55 8	9 55 6

The gradual acceleration of this object during the first two apparitions is well shown on the chart and seems to have been continuous up to 1944–45. The curious 'kink' in its track at about 1946·3 occurred when the p. end of the S.T.Bs., which had been gradually overhauling the prominent feature of that apparition known as the Dark South Tropical Streak, reached conjunction with the p. end of the latter. Both objects underwent a retardation, before the p. end of the S.T.Bs., which had been faint whilst in the longitudes occupied by the Streak, succeeded in breaking away to resume its original rate of progress.

Apparition	Character	Longitude at Opposition	Rotation Period	Rotation Period between Apparitions
		Following End of No. 1 (1f on Chart)		
1939–40	—	—	—	
1940–41	—	—	—	—
1941–42	Z	81°	9h 55m 4s	—
1942–43	B	(110)	—	9h 55m 7s
1943–44	B and Z	124	5	3
1944–45	B	106	5	1
1946	B	154	9 55 6	9
1947	B	(188)	—	9 55 5

The following end of the first section, which did not emerge as a definite feature until 1941–42, showed nothing like the regularity of motion of its p. end; the length of this section must have been continually changing, owing to the variable relative position of its f. end. In 1942–43 there was considerable scattering among the recorded longitudes of the f. end and in 1947 there were two determinations only, both of them, however, quite definite.

Summary of the History of Long-lived S. Temperate Markings
(continued)

Apparition	Character	Longitude at Opposition	Rotation Period	Rotation Period between Apparitions
		Preceding End of No. 2 (2p on Chart)		
1939–40	—	—	—	
1940–41	B	144°	9^h 55^m 14^s	9^h 55^m 6^s
1941–42	Z	175	4	
1942–43	—	—	—	(5)
1943–44	B and Z	206	3	3
1944–45	B	205	4	3
1946	B	213	6	9 55 5
1947	B	238	9 55 8	

The continued identity of this object over the four apparitions 1943–44 to 1947 can hardly be questioned; that the feature of 1940–41 and 1941–42 was the same must be highly probable in view of their courses on the chart and the favourable intermediate period, in spite of the fact that only a single record of it can be found for 1942–43.

Apparition	Character	Longitude at Opposition	Rotation Period	Rotation Period between Apparitions
		Following End of No. 2 (2f on Chart)		
1939–40	—	—	—	
1940–41	—	—	—	
1941–42	Z	189°	9^h 55^m 5^s	9^h 55^m 4^s
1942–43	B and Z	(199)	6	7
1943–44	B and Z	232	7	5
1944–45	B	263	8	7
1946	B	298	7	9 55 6
1947	B	330	9 55 8	

The correctness of the identification of this feature seems to be as well authenticated as in the case of No. 1p.

The steady growth in length of section No. 2 from about 17° of longitude at the beginning of 1943–44 to about 90° at the end of 1947 is apparent on the chart and may also be inferred from a comparison of the rotation periods of its two ends.

Summary of the History of Long-lived S. Temperate Markings
(continued)

Apparition	Character	Longitude at Opposition	Rotation Period	Rotation Period between Apparitions
		Preceding End of No. 3 (3p on Chart)		
1939–40	—	—		
1940–41	Z	(244)°	9^h 55^m 7^s	—
1941–42	B and Z→B	264	7	9^h 55^m 4^s
1942–43	B	262	7	0
1943–44	B	286	6	4
1944–45	B	310	7	8
1946	B	337	5	6
1947	B and Z	6	9 55 9	9 55 5
		Following End of No. 3 (3f on Chart)		
1939–40	—	—	—	
1940–41	B→Z	270°	9^h 55^m 6	—
1941–42	B and Z→B	325	8	9^h 55^m 12^s
1942–43	B	2	6	6
1943–44	Z	21	5	5
1944–45	Z→B	20	4	4
1946	B	44	—	5
1947	B and Z	94	9 55 8	9 55 8

Section No. 3 seems to have become greatly extended in longitude between the end of 1940–41 and the beginning of 1942–43; but if the identification of its p. end is correct, the growth seems to have taken place almost entirely between apparitions, with a halt during the observable part of 1941–42; note the long interim period of the f. end in 1941 and the short one of the p. end in 1942. During 1944–45 it began to contract, reaching a minimum length in 1946 when, however, the f. end was recorded only four times; but it was considerably longer again in 1947. All this depends, of course, on the assumption that the identifications are correct, which seems certain for the p. end from 1942–43 to 1947 and for the f. end from 1940–41 to 1944–45 inclusive and highly probable for the remaining apparitions. In 1939–40 there was a following end lying in a longitude that was too great for the assumption that it was No. 3f to carry much weight; moreover its motion was such that it would have diverged on the chart from the track followed by 3f in 1940–41.

All the longitudes given in these six tables are relative to System II.

124

The South Temperate Zone, which is of course the area between the S.T.B. and the S.S.T.B., is frequently rather a wide zone; but long stretches of it have sometimes been divided into two by a thin grey line, which in places has been strong enough to give rise to doubt as to correct nomenclature. This line was recorded in 1907–08, 1909–10, 1911, 1916–17, 1931–32, 1932–33 and 1938 and has been represented by fragmentary streaks on other occasions; many of these streaks have drifted at the rate of the S.S. Temperate Current. In 1909–10 and 1911 the dusky line provided a remarkable instance of a 'belt' that was not parallel to Jupiter's equator. Starting fairly close and parallel to the S. edge of the S.T.B., it curved southward and, having executed an 'S' bend in the middle of the zone, straightened out again to run close and parallel to the N. edge of the S.S.T.B. On the assumption that the point of inflexion of the bend was drifting with the S.S. Temperate Current, it is highly probable that the same object persisted over the two apparitions.

The brightness of the zone is decidedly variable and on a few occasions, e.g. during a part of 1924, portions of it have been considered brighter than any other region of the disk. Sometimes it has been irregularly shaded, with white spots. The dusky reaches of the zone that were associated with double portions of the S.T.B. from 1939–40 to 1947 have just been described; in 1918–19 a similar shaded strip of the zone, covering ultimately about 140° of longitude, was seen to develop and was described at the time as the S. Temperate Disturbance, owing to its resemblance on a smaller scale to the well-known S. Tropical Disturbance; like the latter it was at one time preceded and followed by bright spots and the two 'disturbances' could be seen on the disk together. The 'S. Temperate Disturbance' moved at the rate of the S.S. Temperate Current. In 1920–21 a somewhat similar disturbed area developed but showed the characteristic form only at its following end, which was also in the S.S. Temperate drift.

The colour of the S. Temperate Zone has usually been described as white or yellow; but in 1928–29 it was affected for a time by a reddish tone that spread over parts of both the S. Tropical and S. Temperate regions.

Addendum

Some time after this chapter had been completed the author had occasion to examine some recent drawings exhibiting three large light oval areas in S. Temperate Zone. The northern edges of these areas, which had sometimes been referred to rather misleadingly as white spots, encroached upon the S. Temperate Belt and at their preceding and following ends there were 'shoulders' that greatly resembled the following and preceding ends respectively of the thicker or double portions of the belt which have already been described.

These white areas, which are well shown in the Frontispiece, in Fox's drawing of Plate V, Fig. 5 and in both the photographs of Plate XII, had been followed by E. J. Reese through three or four apparitions and from the designations he was using to distinguish them, namely FA, BC, DE instead of the more straightforward AB, CD, EF, it seemed clear that A must have been originally assigned to an object that had been registered mentally as the preceding end of a dark feature rather than the following end of a light one. This idea naturally led the author to try to connect the white areas with the spaces between the wider and double portions of the S.T.B. that had been in evidence from 1939 to 1947. The complete success he was able to achieve is now illustrated by Figure 3, of which only the earlier half had originally been prepared. That this chart is now so comprehensive and continuous is due largely to the kind co-operation of various members of the B.A.A. Jupiter Section, in particular the Director, Dr. A. F. Alexander, Mr. W. E. Fox and Mr. D. W. Millar, who have made available to the author many of the Section's original observations and charts.

In order that the continuity of the series may be more readily apparent, the plotting of the apparition shown at the bottom of each of the sloping columns of Figure 3 has been repeated at the top of the ensuing column. At the head of each column are the designations 1p, 2f etc., as employed by the author above, and at the bottom the letters A to F as subsequently assigned by Reese, 1p corresponding to Reese's C. Apart from the fact that 1f(D) was rather poorly observed in 1947 the weakest link in the continuity is to be found in 1943, when the identification of 2p(E) rests upon a single record; that the observation was made by Hargreaves is sufficient evidence of the reality of the object described, but as the region had been well observed throughout the apparition the inference is that the feature was only presented intermittently or that it was generally feeble and insignificant. These two instances are hardly enough to weaken the otherwise overwhelming evidence in favour of the identity of these strong and weak portions of the S. Temperate Belt having persisted continuously over the fourteen apparitions 1941–42 to 1955–56, with the feature 2p(E) dating back to the autumn of 1940 and 3f(B) together with 1p(C) being survivals from 1939.

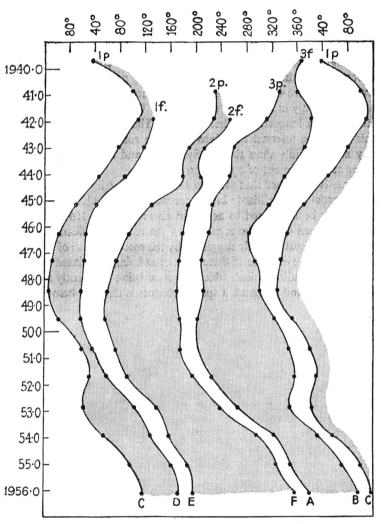

FIGURE 4.—Alternative representation of the behaviour of the S. Temperate features of Figure 3.

127

The question whether the lighter regions should be regarded as spaces between the darker ones or *vice versa* is not quite so arbitrary as it sounds. It will be seen from the chart that except for the shortness of the interval BC in 1939, which very rapidly lengthened, the gaps between the darker regions were wide at first but that they gradually and fairly systematically contracted, until by 1948 they had become three rather short oblong areas, none of which since 1951 has covered more than 30° of longitude. This is particularly well illustrated by Figure 4, which is a chart constructed from the opposition longitudes only in a manner suggested by Reese, who has also been studying the longevity of these objects. The longitudes plotted on this chart are referred to a zero meridian rotating exactly 0°.8 per day more rapidly than that of System II and coinciding with the latter at the moment of opposition in 1952.

There is no record that prior to 1948 any of the lighter spaces had been particularly brilliant; but in recent years the white ovals have frequently been referred to as bright areas or spots. However, that the light areas should be regarded as the really significant features seems to be most strongly suggested by the recent growth of evidence pointing to one or more of these regions as being associated with the outbursts of radio 'noise', which are now being frequently received from Jupiter and to which a special chapter is devoted below.

CHAPTER 14

S.S. TEMPERATE BELT TO
S. POLAR REGION

The S.S. Temperate Belt and Zone
and
The S.S.S. Temperate Belt and Zone

For two reasons it seems best to review the whole of the S.S. Temperate and S.S.S. Temperate regions under a single heading. In the first place, the S.S.S. Temperate Belt is a somewhat ephemeral feature, which cannot always be rigorously classified and to which the name has been applied both when it has been almost a S. component of the S.S.T.B. and when it has been immersed in the S. Polar shading; in the second, there is an absence in the southern hemisphere of the several distinct atmospheric currents that serve in the corresponding northern region as criteria for assigning markings to their proper categories. Indeed, since the B.A.A. Jupiter Section started in 1914 the practice of deriving rotation periods from graphs of the combined longitude determinations of various observers, only two spots ascribed, both of them somewhat doubtfully, to the S.S.S.T.B. have been well enough observed for their rates of drift to be determined; the first, classified in 1928–29 as lying on 'S.S.S.T.B. or belt in S. Polar shading', gave a period of $9^h 55^m 30^s$, while the second, in 1934, whose period was $9^h 55^m 10^s$, though also in high latitude, was clearly under the influence of the S.S. Temperate Current.* Again in 1929–30 there were several grey bands and streaks in these latitudes and one of the spots observed was found to have a rotation period of $9^h 54^m 57^s$, which is short even for the S.S. Temperate Current. During very few apparitions has the S.S.S.T.B. received special mention, though traces of high-latitude belts sometimes appear on the drawings; it must be remembered too that on some of the occasions when the S.S.T.B. has been described as widely double in places, as during the apparitions of 1908–09, 1922 and 1942–43, there has been a fairly wide light 'zone' between the two so-called components.

* See Note D. under the Table of Latitude Measurements on p. 67.

The aspect of the S.S. Temperate Belt is so varied, while the rotation periods of its spots and markings are so comparatively constant, that it is difficult to single out individual apparitions as deserving special comment. It may be noted that the region was considerably disturbed in 1908–09, in 1927–28 and in 1928–29 and that some minute white spots, described as being like satellites in transit, were seen in the belt during 1917–18 and again during 1920–21. The student who is anxious to study in detail the various changes that have taken place in the region must be referred, however, to the Reports of the Jupiter Section of the B.A.A.

Little can be said about the coloration of a feature that is normally rather faint and it is not surprising, therefore, that the S.S.T.B. has generally been considered grey; very occasionally, however, a slightly brownish tinge has been ascribed to it.

References to the S.S. Temperate Zone are confined to a few apparitions. In 1926 a bright strip of the zone was visible, in 1931–32 it was lightly shaded and in 1934 it was sometimes seen to run about halfway round the planet; in 1935 the zone was sometimes prominent, in 1941–42 conspicuous, while in 1942–43 it was very light and wide during the early part of the apparition.

Published references to the S.S.S. Temperate Zone are rather rare, though it is occasionally indicated on a drawing; on 1946 May 12, however, the author was able to see four zones and four belts to the south of the S.T.B.

Undoubtedly one of the most interesting features of this region is the constancy of the rotation period of the S.S. Temperate Current, which has already been mentioned, combined with the comparatively wide range of latitude over which it is operative. In the following table, which gives the mean rotation periods that have been determined for this current, the periods of those S. Temperate objects that have obviously owed their rate of motion to its influence have been included.

Rotation Periods of Spots and Markings in The S.S. Temperate Current

Apparition	Rotation Period	No. of Spots
1887–88	9^h 55^m 3^s	3
1890	7	
1892–93	8	1
1898–99	6	3
1900	7	21
1901	6	21
1902	4	3

Apparition	Rotation Period	No of Spots
1903	8	16
1906–07	11	1
1907–08	5	8
1908–09	5	7
1909–10	6	13
1911	4	6
1914	6	2
1916–17	4	2
1917–18	6	8
1918–19	5	4
1919–20	7	5
1920–21	8	3
1922	11	2
1923	8	2
1925	8	4
1926	8	4
1927–28	10	6
1928–29	8	9
1929–30	7	5
1930–31	10	3
1931–32	10	5
1932–33	10	5
1934	5	6
1935	8(?)	4
1937	8	4
1938	11	3
1939–40	6	5
1940–41	8	7
1941–42	6	13
1942–43	7	11
1943–44	7	14
1944–45	5	15
1946	8	7
1947	8	5
1948	9 55 9	9
Mean	$9^h 55^m 7^s$	

Rotation periods have been determined for so few spots lying beyond the southern limit of the S.S. Temperate Current, that it is not yet known whether there are distinct S.S.S. Temperate and S. Polar Currents, analogous to those found in the northern hemisphere. We may, however, regard it as certain that in the region surrounding the south pole of the planet such features as are occasionally presented are endowed with periods that are considerably longer than and quite distinct from that of the S.S. Temperate Current.

It must be remembered that the northern limit of the shaded area around the south pole is very variable and that the same latitude may be classified as S.S.S. Temperate during one apparition and as in the S. Polar Region during the next; the rotation periods that have been derived from the few available spots in either category are therefore given below in a single short table, in which the letters S.P. in brackets indicate that the features concerned were regarded as S. Polar. The value for 1946 has been roughly assigned by the author as the mean for three objects, whose life was too short for an accurate determination to be possible; it may be some seconds in error but is of the right order of magnitude. Two or three more periods were published during the first decade of this century; but the widely conflicting results of different observers and other considerations do not recommend them as being suitable for reproduction here.

Rotation Periods of S.S.S. Temperate and S. Polar Spots

Apparition	Rotation Period	No. of Spots
1907–08	9^h 55^m 24^s	2 (S.P.)
1928–29	30	1
1934	34	1 (S.P.)
1944–45	27	3
1946	9 55 36	3
Mean	9^h 55^m 30^s	

The South Polar Region

Reference to the table of latitudes on p. 67 will show that on the only occasion for which a determination is given for the N. edge of the S. Polar Region the latitude was considerably higher than any of those that have been measured for the S. edge of the N. Polar Region. The limits are very variable, however, and the general shading frequently extends as far as the S.S. Temperate Belt, as may be gathered from an inspection of drawings and photographs; in February 1943, indeed, the duskiness was sometimes almost continuous from the S. edge of the S.T.B. to the limb.

Most of the infrequent markings displayed here take the form of somewhat darker patches, usually of small area but without definite boundaries, and occasionally a discontinuity in the depth of shading has been seen along a fairly definite meridian, which has not, however, lasted long enough to provide a reliable rotation period.

The region is normally featureless and the only periods available, in which much confidence can be placed, are those given in the foregoing table. Dusky belts in high southern latitude may sometimes be detected, however, as in 1911, 1929–30, 1935 and 1941–42; one or more of these may have been the S.S.S.T.B., but the latitude in 1911 was estimated as $-60°$. In 1922 the area seems to have been more strongly shaded than usual and in 1900, as also in 1912, a dark

132

northern boundary was recorded; in the latter case the drawings suggest that this may have been the S.S.T.B.

From 1893 to 1895 and again from 1903 to 1905 the same ringed or striated appearance was mentioned as was noted in the section describing the N. Polar Region. Neither phenomenon has been recorded during the last forty years; but on one occasion in February 1943, when conditions were superb, Hargreaves, using a 14½-inch reflector, was able to detect that the whole of the region between the S.T.B. and the limb showed a complex pattern of little flecks and streaks.

The colours of the two Polar Regions have been discussed together in the section headed 'Periodicity in Changes of Colour'—p. 65. We have therefore completed our general survey of the visible surface features, leaving only a few objects and phenomena of outstanding interest for specially detailed discussion in the ensuing chapters.

CHAPTER 15

THE GREAT RED SPOT AND THE SOUTH TROPICAL DISTURBANCE

Although the physical characteristics of the Great Red Spot and the South Tropical Disturbance are utterly unalike, yet they shared the same latitudes; and since their motions in longitude differed, sometimes very appreciably, they were repeatedly coming into conjunction, when their interaction was often so remarkable that it would be impossible to discuss them adequately in separate sections of a book, without having continually to be anticipating the one or harking back to the other. Our procedure in this chapter will be, therefore, to begin with the Red Spot and the closely associated bay or 'Hollow' in the southern part of the S. Equatorial Belt, since they are the older and more enduring features; but as soon as it becomes necessary to consider their relation to the Disturbance, the development of the latter will be described in detail, after which we shall return to the Red Spot, tracing its history back as far as it is available and noticing the peculiarities of its motion and a strongly suggested correlation between these and its physical appearance.

Although it was not a new feature, it was in 1878 that attention was first especially directed to the Red Spot, as it was then rising to a prominence that had not previously been recorded; but it was from 1879 to 1882 that it dominated the surface of the planet and achieved immortal fame, not only on account of its being the most conspicuous individual spot ever seen in the matter of size and darkness but also because of its strikingly red colour. Its centre lay a little to the north of the middle of the S. Tropical Zone and its extent in latitude of some 10° was such that it almost touched the S. Temperate Belt, while its northern edge encroached upon the territory of the S. Equatorial Belt. Its shape was that of an elongated oval, the longer axis of which was parallel to the equator and was found by Denning, from a discussion of 252 observations made by 7 different observers, to have had a mean extension in longitude of 33°.7 during the years 1879 to 1882. Its linear dimensions were therefore nearly 40,000 kilometres (25,000 miles) in length and about 13,000 kilometres (8,000 miles) in breadth. The elliptical shape of the Spot, though subject to temporary modifications of detail, is still preserved; but its length, which has varied capriciously, has been less in recent years than that found by Denning. For the fifteen apparitions 1927–28 to 1942–43 the greatest and least extensions in longitude occurred consecutively in 1935 and 1936, when it covered 30° and 21° respectively, while the mean was 25°.5. It is rather remarkable that in 1936, when it was so short, it was probably darker than it had been at any time since 1881; in 1937, when it was almost equally dark, the length was 24°, while in 1927–28, another occasion when it was conspicuous, it was 27°. At the beginning of 1919–20, however, which succeeded an apparition when all but the following end of the Spot had faded to invisibility, it exhibited during its revival the not uncommon attribute of pointed ends and these were so drawn out that for a time its total length was rather more than 40°; subsequently the ends became rounded, but at opposition it still covered 35° of longitude. Variations have also taken place, apparently at random, in the visible breadth of the Spot.

Although the northern portion of the Red Spot commonly occupies latitudes that belong to the S. component of the S. Equatorial Belt, the two features, when visible together as is the normal state of affairs, never overlap one another. Exceptions to this rule are at any rate exceedingly rare; and although a faint linear marking has once or twice been seen traversing the northern portion of the Spot in about the right latitude, it is doubtful whether it can properly be considered to have been the S.E.Bs. Instead, the S. component sweeps northward in a continuous curve around the northern edge of the Spot, from which it is nearly always separated by a narrow light space; the N. component is unaffected, remaining parallel to the equator, and at its most constricted point the whole of the S.E.B. is usually about half as wide as it is in other longitudes, or perhaps a little less. The two components of the belt may remain distinct throughout; but quite commonly they merge together for a short distance where the encroachment of the Spot is deepest. This excavation in the southern part of the S.E.B. is the well-known Red Spot Hollow. Although it has faded on occasions with the rest of the S. component, it is as much a permanent feature of the planet as the Red Spot itself and is usually quite prominent, especially when the Spot is faint. Never,* indeed, as far as we know, have the two objects been invisible simultaneously, which is fortunate in that they have

* Save for a week or two during the great S.E.B. outbreak in 1928.

provided between them a record of the motion of the Red Spot that has been continuous for many years. This implies that the influence, whatever it may be, that is responsible for the presence of the Red Spot is not withdrawn to any marked degree when the spot itself disappears; and in corroboration we have the fact that the depth of the Hollow does not seem to be related to temporary changes in the apparent width of the Spot in latitude.

Not infrequently the south edge of the Red Spot is seen to be in contact with the S. Temperate Belt, sometimes to an extent that would cause the southern part of the ellipse, if it could be traced, to make a considerable incursion on to the belt. Nevertheless, there are few instances of a bay or 'Hollow', analogous to that in the S.E.B., having been seen in the N. edge of the S.T.B. Records of such an appearance do occur, however, the effect being shown, for example, on isolated drawings by J. E. Phocas and by Hargreaves relating to the apparition of 1939–40 and on two made by H. M. Johnson in 1940–41, while a representation by Phillips, dated 1921 April 18, shows the whole of the S.T.B. displaced southwards, though unsymmetrically, in the longitudes occupied by the Spot—see Plate VI, Fig. 5.

In appearance the Red Spot may exhibit any one of three characteristic forms. When it is faint, it is often impossible for the true outline of the Spot to be discerned; and at such times, if it can be seen at all, its presence may be indicated by nothing more than a hazy and somewhat amorphous grey stain on the S. Tropical Zone in the region of the Hollow. When it is an easy object and the whole 'ellipse' can be seen, it may be of comparatively uniform darkness all over, as was the case from 1879 to 1882; or its interior may be very much fainter than its margin, so that its appearance is that of an elongated dark or dusky ring. The latter is by far the more common aspect; and although there are modern instances of the Spot having been of approximately uniform darkness, the ringed form has predominated during the present century. Drawings of the Spot, made prior to 1878, also show the same appearance. Furthermore, in both the ringed and the continuous forms, particularly in the former, of which the appearance of 1919–20, just mentioned, provided an extreme example, the ends of the 'ellipse' have often been pointed, rather like the bows of a ship, a phenomenon that presents great difficulty to anyone who attempts to account for the presence of the Red Spot by invoking the theory of a vortex in Jupiter's atmosphere. On several occasions the major axis of the Spot has been seen to be somewhat inclined to the parallels of latitude and it appears from the records that, whenever this has occurred, the following end has always been the nearer to the planet's equator; such appearances, however, are generally short-lived.

The Red Spot Hollow also exhibits a characteristic form that is alternative to the simple semi-elliptical depression in the S.E.Bs. Curving grey wisps, which are sometimes as dark as the belts they connect, may issue from the ends of the Hollow, bending round across the S.Trop.Z. until they reach the S.T.B. and causing the Hollow to appear as a light oval area, completely surrounded by dusky matter, inside which the outline or traces of the Red Spot itself may or may not be visible; the Spot is rarely other than faint when the Hollow entirely surrounds it. This elliptical form of the Hollow has generally been associated with times when it was in conjunction with the S. Tropical Disturbance and sometimes the dusky shading, preceding and following the apses of the ellipse, has extended for some distance along the S. Tropical Zone. In 1943–44 the Hollow appeared as an almost white oval area in the midst of an extensive dusky region and at the beginning of the apparition was probably brighter than it has been recorded on any other occasion; there was then no trace of the Spot within it, although late in the apparition a small grey fleck was seen. It is of interest to note that the 1943–44 configuration was presented three or four years after the last authenticated observation of either end of the S. Tropical Disturbance. Plate VI provides several examples of the varying aspect c f the Red Spot and Hollow at different stages of their history.

Since the great fading of the Red Spot began towards the end of 1882, after the remarkable years of its prominence, its darkness and general visibility have undergone a number of fluctuations. At first the decline was so steady that before 1890 astronomers had begun to fear that it was doomed to extinction. In 1891, however, a definite revival occurred and it was then more conspicuous than it had been for a number of years. But it soon faded; and, although it was again a fairly easy object in 1893–94, there followed seven apparitions when it was difficult to make it out at all. In 1903–04, 1905–06 and 1906–07, but not in 1904–05, it was very plainly seen and, after a further period of faintness, it was again a distinct and well-defined ellipse in 1914. Three apparitions later, in 1917–18, there was another marked revival. In 1919–20, after the fading of the other S. Tropical markings, including the S. Tropical Disturbance, the S. component of the S.E.B. and even the Hollow, the Red Spot became quite conspicuous; with their return to visibility, it lost its prominence but remained a fairly easy object during the next apparition, after which it faded.

In 1926 the familiar S. Tropical features again disappeared and remained practically invisible until 1928–29; this period was remarkable for a fresh revival of the Red Spot, which in 1927–28 became outstandingly conspicuous. After a rapid extinction in 1928–29, during which it appeared to be swamped by the returning S. Tropical detail, there followed a rather slow fluctuation, which culminated with a moderate maximum in 1931–32; and then, during yet another fading of the S.E.Bs. and the Disturbance, it achieved in 1936 and 1937 a prominence that must have been little inferior, except as regards its extension in longitude, to that of the years around 1880. The return of the missing S. Tropical features once more reduced it to insignificance, so that in 1938 it was practically invisible. Since then there have been further fluctuations of a normal character but nothing spectacular has occurred.

Both in 1927–28 and in 1936, when the Red Spot was so outstandingly prominent, the whole region between the N. component of the S. Equatorial Belt and the S. Temperate Belt was brilliantly white and it was also white in 1919–20. On these occasions the S.E.Bs. and other S. Tropical markings seem to have been obliterated by the condensation of a vast obscuring cloud, which for some reason was unable to form above the Spot itself. Supporting evidence is contributed by the extreme brilliance of this zone on the photographs taken in ulta-violet and violet light by W. H. Wright in 1927–28. In the drawings of 1880 and 1881 also the S.E.Bs. is missing and the region is shown lighter than the rest of the disk; so it is fairly safe to infer that similar conditions prevailed. The great darkness of the Red Spot on all these occasions must, therefore, have owed something to the effect of contrast with the brightness of the background against which it appeared; for we know that contrast alone is responsible for converting the appearance of Satellite IV, when in transit, from a bright object just after ingress to an almost perfectly black spot when it is projected against the brighter central parts of the disk. Further and more detailed reference to the darkness of the Red Spot will be made at a later stage in this book, when speculations regarding its nature and behaviour will be under review.

The author would willingly forgo his duty to discuss the actual redness of the Great Red Spot, especially as he himself has never noticed anything more striking in its coloration than a faint tendency to pinkness. He is quite incapable of imagining what its colour must have resembled during the notable years of 1879 to 1881 and can only state his belief that the lavish application of bright red pigments to the numerous tinted drawings that are extant conveys a very exaggerated impression of its true aspect. Lest it should be suggested that he is himself colour-blind, he may perhaps be permitted to state that in ordinary life he is at least normally sensitive to slight variations of tint and is probably rather more than ordinarily red-sensitive, since he habitually considers that Arcturus is visually a brighter star than Vega. One of the difficulties confronting anyone who attempts to collate the colour estimates of others is that the term 'red' seems to be applied by some individuals to any shade of which the effective wavelength is greater than that of the 'D' lines. Some years ago the author had occasion to ask a well-known observer whether the expression 'a full fiery red', which he had used in the description of some feature on Jupiter, referred to the flame or to the heart of a fire. The reply was that the comparison was intended to be with the flame; but a flame is surely orange-yellow, not red. However, in the following paragraph an attempt has been made to indicate the varying degrees of redness that are found in the records to have been attributed to the Spot since 1882.

Naturally, when the Spot has been very faint, it has conveyed little impression of colour. In 1886–87, although Terby described it as having lost its characteristic tint, Williams, whose eye seems to have been particularly sensitive to faint traces of red in the Jovian markings, always found a distinct reddish or pinkish tinge in the Spot and in the following apparition its appearance to him was similar; in 1896–97 we find him still referring to it in much the same terms. In 1909–10 E. M. Antoniadi described the colour as intensely pink, while E. Hawks called it a full grey although there was a suggestion of warmth about it on several occasions that was not so pronounced that it could be classed as red or even pink; Phillips also found it grey. The last two employed reflectors of 18 and $12\frac{1}{4}$ inches aperture respectively, while Antoniadi was presumably using the 33-inch refractor of the Meudon Observatory, with which his drawings of that apparition were made and with which he again saw it intensely pink in 1911. In 1926 and 1927–28, when the Spot was so prominent, it was described as having recovered much of its warm tone; in 1926 C. F. Du Martheray found that the edges were deep red, particularly at the following end, and in 1927–28 F. Sargent called it 'brick-red' or 'pale carrot'. On Wright's photographs, taken during the latter apparition, the images of the object are very dark in light of short wavelengths but almost invisible in the red and infra-red—Plate I, Figs. 5 and 6; this, of course, implies that the light scattered by it was rich in the longer wavelengths and confirms the visual impressions of warmth in its coloration, but may also have been in part an effect of contrast, since we have already noted that its light background was deficient in the longer radiations. In 1934 two observers independently called it salmon pink and Williams, who employed a scale of redness from 1 to 10, made it about 3 on the average. During the great prominence of the Spot in 1936 and 1937 most observers found it decidedly reddish in tone and Williams recorded that the red colour was conspicuous in 1936; in 1937 he sometimes called it pinky red and generally allotted to it the number 6 on his scale, never less than 5. In 1938, when there was some doubt as to whether the dark elliptical outline that remained after the return of the S.E.Bs. was the Red Spot itself or the boundary of the Hollow, Phillips wrote of its light interior that, except for a bright area near its preceding end, it was 'pinky red—perhaps carmine—a beautiful colour'. It is interesting to find that as early as 1877 a group of observers in New South Wales had nicknamed the Red Spot 'The Pink Fish'.

Measurements of the latitude of the centre of the Red Spot were made with a filar micrometer by Professor G. W. Hough at the Dearborn observatory from 1879 to 1882 and were published, after reduction to seconds of arc, presumably referred to Jupiter's equator, at the mean distance of the planet from the Earth. When these are further reduced to zenographical latitude, the following figures are obtained:

Apparition	Zenographical Latitude
1879–80	−23°.6
1880–81	−24·3
1881–82	−25·2

In recent years few direct measurements have been made of the latitude of the Spot. Phillips found it more practicable to measure its position on his drawings in relation to the S. Temperate Belt and to the N. and S. edges of the S. Equatorial Belt, the position of each of these reference lines having been obtained by means of the micrometer. The following table gives his results, determined by this method, from 1908–09 to 1929–30.

The author has measured two more recent drawings of his own, on which the S.T.B. was very accurately placed with the aid of the micrometer, and has found zenographical latitudes of a little less than −21° in 1935, when the S. edge of the Spot was well clear of the S.T.B. and extending deeply into the Hollow, and about −25° in 1943, when it was hard up against the S.T.B., which was perfectly straight, for almost the whole of its length. The only published value of a direct micrometrical measurement made by Phillips is dated 1936 June 9 and gives the remarkable latitude of $-19°7$. This is less than $1°$ south of the mean latitude of the S. edge of the S.E.B. and is so much farther north than the position of the Spot on drawings he made in March and August of that year, that one is suspicious of an error, possibly in the reading of a division on the micrometer head.

On the whole, however, the departures of the figures given for individual apparitions from the mean of all the above determinations seem to be large enough to justify the assertion that the latitude of the centre of the Red Spot is subject to variations with a range of at

Apparition	Zenographical Latitude	Apparition	Zenographical Latitude
1908–09	$-20°4$	1919–20	$-20°4$
1909–10	$-22·0$	1920–21	$-22·3$
1911	$-22·5$	1922	$-21·4$
1912	$-22·8$	1923	—
1913	$-22·2$	1924	$-22·9$
1914	$-22·3$	1925	$-23·5$
1915	$-22·3$	1926	$-23·0$
1916–17	$-21·8$	1927–28	$-21·6$
1917–18	$-20·0$	1928–29	$-21·5$
1918–19	$-20·8$	1929–30	$-21·0$

Mean $\quad -21°8$

Impressive though the run of the figures is, the accidental errors of measurement seem likely to have been large enough to obscure the small systematic oscillation in a period of about twelve years which was suggested by Phillips.

least two or three degrees. After all, it is not surprising that the latitude of an atmospheric configuration should vary; the really astonishing thing is that the changes of latitude are not accompanied by violent alterations in rotation period. The displacement of any material substance through only $1°$ of latitude from $-23°$ to $-22°$ would bring about a lengthening of its radius of rotation about the planet's axis of rather more than $0·7$ per cent; and for its angular momentum to be conserved this would have to be accompanied by an increase in period of about $1·5$ per cent, or nearly 9 minutes! Yet over the apparitions of 1926 to 1929–30, when the measured latitude of the Spot changed from $-23°0$ to $-21°0$, its rotation period varied by as little as $1·3$ seconds. The extreme range in the rotation periods that have been recorded for the Red Spot is 12 seconds and the changes have by no means always been in the right sense to correspond with those in its latitude. But before we discuss further the motion of the Red Spot, we must turn to the consideration of that remarkable feature known as the South Tropical Disturbance.

During the apparitions of 1900 and 1901 a series of dark humps could be seen, projecting somewhat from the S. edge of the S. Equatorial Belt into the S. Tropical Zone. One of these was especially prominent early in 1901 and on February 28 Molesworth recorded that it had become connected, by a grey ligament across the zone, with a dark streak on the S. Temperate Belt. This dusky ligament became rapidly distended in longitude and brilliant white spots developed at its preceding and following ends. It was to the shaded part of the S. Tropical Zone between these white spots that the name 'South Tropical Disturbance' was given. Unfortunately the record of its exact longitude on 1901 February 28 appears to be lost; but a satisfactory rotation period of $9^h 55^m 20^s$ was obtained for the Disturbance during the remainder of that apparition, from which a reasonably trustworthy extrapolation places it close to 313° in System II on that date. It may be noted that this was at least 90° from the longitude of the Red Spot, which at that time was about 47°.

Since the rotation period of the Red Spot in 1901 was $9^h 55^m 41^s$, the new feature, if sufficiently long-lived, was clearly destined to overtake it in due course. In spite of an initial handicap of some 270°, the Disturbance, which had continued to exhibit increasing extension in longitude, accomplished this feat in June 1902, when its preceding end reached the following end of the Red Spot Hollow. Interest was naturally centred upon the manner in which the two objects would react upon each other; and, as so often happens under such circumstances, what actually occurred was unexpected. During the period of at least six weeks, which should have been required for the p. end of the Disturbance to pass from one end of the Hollow to the other, and indeed during the whole of the conjunction there was no sign whatever of any encroachment upon the region; instead, within a few days of its arrival at the f. end of the Hollow, a facsimile of the p. end of the Disturbance was seen by Molesworth to be forming at the other end of the Hollow, with the result that the Hollow itself, having become completely surrounded by the dusky shading, assumed the now familiar form of a light ellipse on a grey background, within which the Red Spot could be distinctly discerned by those possessing adequate instruments, though it was very much fainter than the dusky region of the Disturbance. The new development at the p. end of the Hollow proved to be a true p. end of the Disturbance and it drew away from the Red Spot at approximately the same rate as that which it had shown when approaching it prior to conjunction. Its rapid leap across the confines of the Red Spot resulted in the addition of nearly the whole of the length of the

Hollow to that of the Disturbance, which then totalled about 90°. The duration of this first conjunction was a little more than three months, the end occurring in 1902 September.

The second conjunction occurred in 1904, the length of the Disturbance having contracted during the interval to reach a minimum of 35° in 1903. In 1904 the maximum length attained was 80°, which included the length of the Hollow at conjunction, but it had fallen again to only 30° at the end of the year. The motion of the f. end of the Disturbance just before and after the end of the second conjunction indicates approximately the same date for its arrival at the f. end of the Hollow and its departure from the p. end; although the actual process does not seem to have been recorded, we are led to the conclusion that its transference across the Hollow was practically instantaneous and similar to that which was observed in the case of the p. end in 1902. At the beginning of the third conjunction in 1906 the behaviour of the p. end of the Disturbance was carefully studied. It reached the f. end of the Hollow on February 25; but for some weeks nothing was seen of it on the p. side of the Hollow. At the beginning of April, however, it became evident that the S. Tropical Zone in this region was growing darker, although it was not until the middle of the month that it was possible to identify any particular longitude with that of the p. end of the Disturbance. It was then seen to have advanced by an amount which, on extrapolation from its motion, gave March 11 as the date of its conjunction with the p. end of the Hollow. Thus its passage through the longitudes occupied by the Red Spot region, which would have taken nearly three months at its normal rate of progress, must have been accomplished in a matter of fourteen days. The end of the third conjunction did not occur until after the beginning of the next apparition, when the transference of the f. end of the Disturbance from one side of the Hollow to the other seemingly occupied only twelve days, though some of the observations were a little uncertain.

The indefinite manner in which the p. end of the Disturbance materialised after the beginning of the third conjunction was exhibited on several subsequent occasions. In all there were nine conjunctions and perhaps the beginning of a tenth; but to describe them all in detail here would occupy too much space. The accompanying table, which is an augmented abstract of one given by Phillips in the Twenty-ninth Report of the B.A.A. Jupiter Section, summarises well the interaction of the Red Spot and the Disturbance at such times and some of the points it brings out will be referred to in a moment. Meanwhile it must be mentioned that in 1914, at the end of the sixth conjunction, the f. end of the Disturbance was observed to occupy a little more than six weeks in passing over the 37° then covered by the Hollow; on that occasion, therefore, although its

Interaction of the Great Red Spot and the South Tropical Disturbance

Apparition	Rotation Period of Red Spot	Rotation Period of Disturbance		Conjunction	Length of Disturbance
		p. end	f. end		
	9ʰ 55ᵐ+	9ʰ 55ᵐ+	9ʰ 55ᵐ+		
1901	41ˢ	20ˢ	20ˢ		A few degrees
1902	39	16	16	1st	Max. about 90°
1903	41	centre 21			Min. 35°
1904–05	40	21	23	2nd	Max. 80°; later 30°
1905–06	41	19	23	3rd	65°
1906–07	42	22	22		57°–45°
1907–08	41	17	22	4th	95°–100°
1908–09	42	20	26		57°–70°
1909–10	38	21	20	5th	108°
1911	38	20	30		115°
1912	39	27	26		65°
1913	35	28	24	6th	Max. at least 140°
1914	36	28	—		109°
1915	37	31	28		92°
1916–17	36	28	28	7th	119°
1917–18	34	25	31		167°
1918–19	37	30	30		Max. 190°+
1919–20	35	33	29		136°
1920–21	38	30	33		134°
1922	39	31	32	8th (partial)	151°
1923	37	—	32		160°±
1924	32	34	41		145°
1925	33	34	39		150°
1926	36	—	—		Invisible
1927–28	38	—	—		Invisible
1928–29	38	15	41		220°
1929–30	38	32	36		214°
1930–31	39	33	37		202°
1931–32	38	35	37		213°
1932–33	39	37	37	9th	227°
1934	39	38	39		224°
1935	39	38	40		230°
1936	41	—	—		Invisible
1937	42	—	—		Invisible
1938	42	33	—		Uncertain
1939–40	44	38	34	Beginning of 10th?	207°

motion was accelerated, there was no sudden leap from one end of the Hollow to the other. Some aspects of the eighth and ninth conjunctions will be described after a further short review of the behaviour of the Disturbance in general.

It will be seen from the Table that the length of the Disturbance was subject, during most of its history, to marked fluctuations but that these were superposed on a general tendency for it to grow longer, which culminated in a maximum extension of 230°, or nearly two-thirds of the circumference of the S. Tropical Zone, in 1935. As time went on, the central regions, which had been almost uniformly dark in the early stages, became somewhat irregularly shaded and ultimately faded to such an extent, that towards the close of its career only the characteristic forms of its two ends remained as identification marks. These were last seen, faintly but definitely, during the apparition of 1939–40, since when no claim regarding the location of either has been substantiated. The bright spots, one preceding the p. end, the other following the f. end, were often conspicuous during the early years of its history; but later, when visible at all, they had lost their brilliance. The two ends of the Disturbance, though subject to some variations of form, maintained in general their concavity towards the positions of these spots, so that the aspect of the whole, imprisoned between the S.E.B. and the S.T.B., could be compared with that of a column of fluid in a thin tube, complete with a concave 'meniscus' at either end. Plate VII shows several characteristic views of the Disturbance and its two ends.

Referring again to the Table, we notice that the rotation periods of the Disturbance and the Red Spot, which at first differed by some twenty seconds, gradually approached one another until, during the latter part of the long ninth conjunction, they were almost equal. The effect, or at any rate the concomitant, of conjunctions with the Red Spot seems to have been an acceleration of the latter; for at each conjunction before the ninth there was a tendency, sometimes marked, for its rotation period to shorten. During most of the ninth conjunction, however, the two periods were so nearly equal that the effect of their interaction, if any, could not have been expected to be large. The periods of only $9^h 55^m 16^s$ shown by both ends of the Disturbance in 1902 and of $9^h 55^m 17^s$ and $9^h 55^m 15^s$ by the p. end in 1907–08 and 1928–29 respectively are accounted for primarily by the fact that the very rapid passages of these features across the longitudes occupied by the Hollow are included in their mean motions. Although this phenomenon had not been actually recorded since the fourth conjunction, at the beginning of the ninth the p. end of the Disturbance was seen to be appearing at the p. end of the Hollow within a few days of its arrival at the f. end!

During the earlier conjunctions the effect of the Red Spot on the motion of the Disturbance was in general such that, whenever either end of the Disturbance was approaching it, that end was somewhat accelerated, while after passage across the Hollow each end of the Disturbance would be retarded, almost as though some attractive force had been at work.

The two factors that combined in causing the conjunctions of the Red Spot and the S. Tropical Disturbance to occur at progressively longer intervals and at the same time to become gradually more protracted were the approach to equality of their rotation periods and the increase in length of the Disturbance. To this rule, however, the eighth conjunction provided a remarkable exception. Not only was it of short duration but it was only partial. The p. end of the Disturbance reached the f. end of the Hollow shortly after the beginning of the apparition of 1923 but had not been seen at the other end before the close of the apparition. When the planet was first seen again in 1924, the state of affairs was similar; but the acceleration of the Red Spot, usually associated with such times, had attained such a magnitude, that its rotation period had become shorter than that of the Disturbance for the first time and before long it had drawn clear ahead, the p. end of the Disturbance being once more visible a few degrees following the f. end of the Hollow. This gap had widened to about 40° at opposition in 1925. Between the oppositions of 1923 and 1924 the System II longitude of the Red Spot had decreased by 58° and between those of 1924 and 1925 by as much as 82°, the mean rotation period of $9^h 55^m 32^s$ in 1924 being the shortest that has been recorded, since it first became a conspicuous object. Exactly what happened during the next three years is unknown, as the Disturbance, the S. component of the S. Equatorial Belt and the Red Spot Hollow, but not the Red Spot itself, disappeared, having been apparently obscured by bright overlying cloud. After the spectacular return of these features during the apparition of 1928–29 it was found, however, that the Disturbance was again rotating in a shorter period than the Red Spot and that its length had increased to more than 200°. When it was first recovered in August 1928, the p. end was following the f. end of the Hollow by about 20° and conjunction began in November.

During 1936 and 1937, before the end of the long ninth conjunction, the S. Tropical markings were again obliterated, as they had been from 1926 to 1928. Their return, while Jupiter was near the Sun between the apparitions of 1937 and 1938, revealed only the p. end of the Disturbance; but in 1939–40 the f. end was also definitely seen and disclosed the fact that conjunction was over at last. Meanwhile the p. end, though fading, was again drawing in towards the Red Spot and the beginning of a tenth conjunction; and on 1939 December 27 Phillips wrote: 'The S. Tropical Zone is lightly shaded from the f. end of the Red Spot onwards. Has the Disturbance overtaken the Red Spot? I am inclined to think it has.' And that is the last definite record we have of an object that had been a familiar 'landmark' on Jupiter for nearly forty years.

It is possible that the appearance of the S. Tropical Disturbance in 1901 was not the first manifestation of such a phenomenon. Two or three of the older drawings show markings, whose appearance is remarkably suggestive of the p. end, and at least one contains a hint of the f. end. Among Rev. W. R. Dawes' drawings of 1857, reproduced in *M.N.* 18, 50, we find possible p. ends, nearly up to the C.M. at $6^h 43^m$ on November 28 and about central at $7^h 30^m$ on December 5, a possible f. end being also shown on November 28, just past the C.M. at $8^h 29^m$. Moreover, a search through the original drawings in the R.A.S. collection has revealed that in 1874 E. B. Knobel indicated a 'shoulder' in the S.E.Bs., which might well have been a p. end of the Disturbance, at 65° System II on March 7 and again at 64° on March 24, while on the latter date another similar configuration appears at about 92°, without however any indication of the f. end. But by far the most striking of the R.A.S. drawings is one executed by H. Schwabe, of sun-spot fame, at $10\frac{1}{4}$ hours on 1859 November 13, a copy of which is reproduced here—Plate V, Fig. 6; it will probably be agreed that this drawing depicts what might well be a representation of the p. end of the Disturbance at almost any time between 1901 and 1939.

In spite of these resemblances, there is no evidence that any of the above were more than transient features. Whether the Disturbance will be seen again the future alone can decide; and if a similar object were to appear, there would be no means now of identifying it with the original of 1901 to 1939. The 'Dark South Tropical Streaks' of 1941–42 and 1946 simulated it in so many respects that on both occasions it was thought at first that the Disturbance had not only reappeared but had become rejuvenated. These two notable objects will be described in the next chapter; and we shall once more find the South Tropical Disturbance playing an important role in the chapter assigned to the 'Circulating Current'.

We return now to the early history of the Great Red Spot and to a more detailed discussion of its motion, bearing in mind that between 1901 and 1940 conjunctions with the S. Tropical Disturbance may not have been without their influence upon the latter. Its history has been traced back, in a paper (*M.N.*, 59, 574) published by W. F. Denning in 1899, to a representation of the Hollow on a drawing executed by Schwabe on 1831 September 5. This drawing is of paramount importance, since it constitutes the earliest definite record we possess and establishes the fact that the Red Spot, or at any rate the Hollow, has been in existence for over 120 years. Denning also made the suggestion that the somewhat smaller dark object, whose discovery upon the S. Tropical Zone in 1664 is generally attributed to Hooke, was a very much earlier manifestation of the Red Spot itself. As has already been mentioned, this spot was recorded intermittently from 1664 to 1713, so it was at least more permanent than the majority of Jovian markings. The author has recently consulted Stanley Williams' original notes of 1936 and 1937 in the hope that he might find some comparison of the aspect of the Spot at that time with that which Williams himself had beheld in 1880; although the reference sought was not forthcoming, there was a most interesting note to the effect that its appearance in 1936 reminded him strongly of a drawing he had seen of Hooke's spot, made by Cassini in 1665. If, as seems highly probable, the identification is correct, then the historical life of the Red Spot is approaching 300 years, a truly venerable age for a feature that is manifested only in the planet's atmosphere.

That what we observe is not related to some landmark on the solid surface of a uniformly rotating planet is an immediate inference from the changes that have taken place in its rate of rotation, which can be followed by referring to the following table, where all the mean rotation periods available to the author are given; doubtless several others have also been determined.

Rotation Periods of the Great Red Spot

Apparition	Rotation Period	Apparition	Rotation Period
1872–73	$9^h\ 55^m\ 31^s$	1914	$9^h\ 55^m\ 36^s$
1879–80	34	1915	37
1880–81	36	1916–17	36
1881–82	38	1917–18	34
1882–83	39	1918–19	37
1883–84	39	1919–20	35
1884–85	39	1920–21	38

Apparition	Rotation Period	Apparition	Rotation Period
1885–86	41	1922	39
1886–87	41	1923	37
1888	41	1924	32
1890	40	1925	33
1891	42	1926	36
1892–93	41	1927–28	38
1893–94	41	1928–29	38
1894–95	41	1929–30	38
1895–96	41	1930–31	39
1896–97	42	1931–32	38
1897–98	42	1932–33	39
1898–99	42	1934	39
1900	42	1935	39
1901	41	1936	41
1902	39	1937	42
1903–04	41	1938	42
1904–05	40	1939–40	44
1905–06	41	1940–41	41
1906–07	42	1941–42	40
1907–08	41	1942–43	41
1908–09	42	1943–44	42
1909–10	38	1944–45	44
1911	38	1946	43
1912	39	1947	42
1913	9 55 35	1948	9 55 43

Mean Rotation Period $9^h 55^m 37^s.6$

As a supplement to the above table we have the following mean values of the rotation period, computed by Denning from records of longitude determinations, over a number of consecutive intervals between the years 1831 and 1878.

Mean Rotation Periods of the Great Red Spot during the years 1831 to 1878

Years	Mean Rotation Period	Years	Mean Rotation Period
1831–32	$9^h 55^m 32^s$	1857–58	$9^h 55^m 37^s$
1832–40	34	1858–60	38
1840–45	35	1860–63	36
1845–50	35	1863–67	34
1850–51	36	1867–69	34
1851–52	35	1869–73	34
1852–55	36	1873–76	34
1855–56	37	1876–78	9 55 33
1856–57	9 55 38		

To these may be added for comparison two determinations made by Cassini of the mean rotation period of Hooke's Spot.

Mean Rotation Periods of Hooke's Spot of 1664 to 1713

Years	Mean Rotation Period
1664–66	$9^h\ 55^m\ 59^s$
1666–72	$9\ \ 55\ \ 54$

The period is longer than any found for the true Red Spot but appears to have been shortening. The author is uncertain whether the observations of Cassini were subsequently corrected for light-time, the effect of which was discovered by Roemer in 1675; but its neglect is unlikely to have affected the periods deduced by more than a second or two.

It is probably more instructive, however, to study the drift of the Spot in longitude. If an object is stationary, relative to a uniformly rotating planet, the graph of its longitude against the time, measured from any arbitrary but uniformly rotating meridian, such as the zero of System II, will clearly be a straight line. That the graph of the longitude of the Red Spot, plotted in System II, is far from linear is immediately evident from an inspection of the next Table.

In the first column are given the years in which opposition occurred and in the second the number of the opposition, the epoch or zero opposition having been chosen as that of 1894, when the System II longitude of the Spot was 0°. We see from the third column, in which the System II longitudes at opposition are tabulated, that they range from $+2588°$ in 1831 to $-941°$ in 1937 over a total of 3529° or nearly ten complete rotations; which suggests that, if the sole purpose of System II had been to supply a reference meridian for the Red Spot, a more convenient value for its rotation period might have been chosen. But we also recognise at a glance that there were turning points in the motion, notably in 1890 and in 1901; and these ensure that, although we may be able to find a more suitable uniformly rotating meridian from which to measure the longitudes, the graph can never become a straight line.

It is of interest, however, to discover the uniform system that is most effective in reducing the wanderings of the Red Spot in longitude. The choice will depend to some extent on the exact conditions that are to be fulfilled; but one of the ways of defining such a zero meridian is that the sum of the squares of the longitudes measured from it must be a minimum.

If λ_0 be the System II longitude of the new zero meridian at the epoch, λ_2 and λ being the longitudes of the Red Spot in System II and in the new system respectively, then

$$\lambda = \lambda_2 - (\lambda_0 + \alpha t)$$

150

Drift of the Great Red Spot in Longitude 1831 to 1952

Year	t	λ_2	λ	Year	t	λ_2	λ
1831	−58	+2588	+664	1910	+14	+8	+144
1832	57	2500	604	1911	15	−32	133
1840	50	2096	401	1912	16	56	138
1845	45	1829	277	1913	17	99	123
1850	41	1625	187	1914	18	156	95
1851	40	1577	168	1915	19	187	92
1852	39	1526	146	1916	20	230	78
1855	36	1389	94	1917	21	281	56
1856	35	1355	89	1919	22	350	+15
1857	34	1326	89	1920	23	398	−4
1858	33	1294	85	1921	24	436	13
1860	32	1276	96	1922	25	464	13
1863	29	1129	+35	1923	26	490	10
1864	28	1062	−4	1924	27	548	40
1867	25	887	93	1925	28	630	93
1869	23	765	158	1926	29	699	133
1873	20	584	253	1927	30	736	142
1876	17	397	354	1928	31	765	142
1878	15	256	438	1929	32	796	144
1880	13	128	508	1931	33	823	143
1884	10	35	515	1932	34	851	142
1887	7	+13	452	1933	35	877	140
1889	5	−6	413	1934	36	900	134
1891	3	6	356	1935	37	921	126
1892	2	4	326	1936	38	938	115
1893	−1	−2	295	1937	39	941	89
1894	0	0	264	1938	40	933	53
1896	+1	+7	229	1939	41	914	−5
1897	2	15	192	1940	42	907	+31
1898	3	22	156	1941	43	911	55
1899	4	31	119	1943	44	908	87
1900	5	40	81	1944	45	897	127
1901	6	46	47	1945	46	868	184
1902	7	42	22	1946	47	863	218
1903	8	34	−1	1947	48	850	259
1904	9	25	+18	1948	49	846	292
1905	10	28	50	1949	50	839	328
1906	11	18	69	1950	51	830	365
1908	12	23	102	1951	52	824	400
1909	+13	+17	+125	1952	+53	−809	+444

α being a constant and t the time; and we wish therefore to determine the values of λ_0 and α that correspond to the minimum value of $\Sigma\,(\lambda_2-\lambda_0-\alpha t)^2$.

The solution obtained by giving to λ_2 the values taken from column 3 of the Table and to t those from column 2 is

$$\lambda_0 = +264°\!.3; \quad \alpha = -28°\!.62$$

The unit of t has in effect been made equal to the mean interval between oppositions, or 398·88 days, while the figures in column 2 refer to the actual dates of opposition; but the drift of the Red Spot is so small during the few days by which the mean and the true dates differ, that it can be neglected.

The values of λ, derived from the above expression, are given in column 4 of the Table and in Fig. 5 they are shown plotted against the time, as represented by the figures in column 2. Although column 4 is a great improvement on column 3, we see that the wanderings of the Red Spot since 1831 still range through 1179° from +664° to −515° and then back by a sinuous route to +444°. If the zero of System II were plotted on this diagram, it would follow a straight line, passing through −264° at $t=0$ and keeping fairly close to the path of the Red Spot until it reached −165° at $t=+15$.

The rotation period of the zero meridian of the new system, $\lambda=0°$, is $9^h 55^m 37\overset{s}{\cdot}58$, which is in complete agreement with the mean, given in the Table on page 145 for the interval from 1872 to 1948.

It is possible to choose a reference system in which the total range in longitude is reduced to 1080°, or exactly three revolutions, but no less. The track of its zero meridian in Fig. 5 would lie parallel to the line joining the positions of the Red Spot at $t=-58$ and $t=+53$ and its rotation period would be $9^h 55^m 37\overset{s}{\cdot}48$. As, however, the longitude of the Spot at the time of writing is still increasing, this would probably be only a temporary subterfuge.

The results of the above investigation demonstrate conclusively that the position of the Red Spot on the planet cannot bear any relation to a source that maintains a fixed position on the surface of a uniformly rotating solid body.

The four points, marked A, B, C, D on the curve in Fig. 5, are of outstanding interest. Each is associated with a positive acceleration of the Red Spot and three of them mark the beginning of such an acceleration. The curvature of the graph, that gives rise to the 'hairpin' bend at about $t=-12$, starts at A, where $t=-13$; and this corresponds to the year 1880, when the Spot was so conspicuous. The point B, at $t=+22$, indicates the initial stage of another, somewhat less pronounced, acceleration during the apparition of 1919–20, when the Spot again became an easy object. At C, where $t=+29$, there is yet another rather abrupt acceleration; this occurred during the apparition of 1926, when the Spot became an outstanding object. The year 1936, when the Red Spot was darker than it had been at any time since 1882, is represented by the point D at $t=+38$; although it does not occur at the beginning, it marks the time at

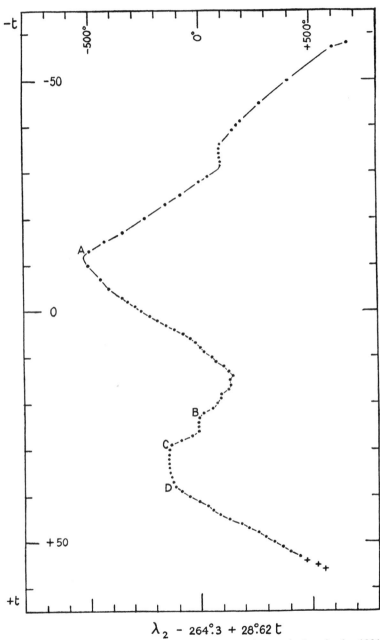

FIGURE 5.—The wanderings of the Great Red Spot in longitude, 1831 to 1955.

which an aready existing acceleration showed a very rapid increase. The reader should not be misled into thinking that the greatest acceleration corresponds to the greatest curvature of the graph, as the latter depends largely upon the reference system chosen.

It will be noticed that no reference has been made to the acceleration that clearly began at about $t = -36$ in the year 1855. The records relating to that period are not very numerous; and it seems not unlikely that the Spot could have achieved a darkness, at least as great as that of 1919–20, without having evoked any special comment.

It appears, therefore, that there is a *prima facie* case for suspecting a physical relation between the darkness and the acceleration of the Red Spot; the interdependence of these qualities will be examined in greater detail in the theoretical part of this book.

Many further references will be made to the Great Red Spot in connection with other phenomena; but it is time to take temporary leave of this historic feature and to turn our attention to other remarkable events that have been witnessed in that astonishing region, the S. Tropical Zone.

Addendum

It is seldom that anything of importance is revealed to the casual observer; yet on 1955 September 19, when the author out of idle curiosity pointed a 3-inch telescope at Jupiter low down in the morning twilight, he perceived at once that in a longitude somewhat preceding that of the central meridian the S. Equatorial Belt and the S. Temperate Belt were connected by some sort of dusky ligament, which must have lain close to 245° in System II. The extent of this object in longitude was estimated to be about 5° and the whole appearance immediately called to mind the description of Molesworth's original observation of the S. Tropical Disturbance in 1901.

The extension of the new feature in longitude soon began to increase, the initial rotation period of its preceding end being about $9^h 55^m 15^s$ while that of the following end averaged $9^h 55^m 36^s$ until the end of 1955. This corresponded at first to a rate of growth equal to some 16° in thirty days. From about November 10, however, the motion of the p. end became retarded, its period increasing to approximate equality with that of the f. end; the growth of the object was thus arrested until about December 16, when the p. end started to draw away again in a period of $9^h 55^m 25^s$. The increase in length practically ceased again when, at the beginning of January 1956, the f. end became accelerated to a period of about $9^h 55^m 26^s$. Early in

March both ends slowed down, the p. end to a period of about $9^h 55^m 30^s$ and the f. end to one of a second or two longer, the extension thus continuing slowly to increase until it reached a maximum of some 42° of longitude shortly before the end of the apparition.

The above figures are provisional only and are the combined result of a study by the author of his own rather meagre observations and of correspondence with Mr. W. E. Fox, who has kindly communicated in advance of publication his preliminary interpretation of the B.A.A. longitude charts.

Of particular interest to the author is the question whether this latest disturbance in the S. Tropical Zone is more closely analogous to the old South Tropical Disturbance or to the Dark South Tropical Streaks of 1941–42 and 1946. The former ended its long career by simply fading imperceptibly away; the latter objects maintained their darkness but disappeared by contracting in longitude.

At the beginning of the apparition of 1956–57 the S. component of the S.E.B. was seen to have faded almost to vanishing point and the new feature, presumably under the same influence, had been reduced to patches of very indefinite grey shading, the ends of which were often most difficult to locate with accuracy. Now it will be remembered that on two or three occasions the old S. Tropical Disturbance suffered complete though temporary extinction, together with the S.E.Bs. and the Red Spot Hollow, and the progress of events during the autumn of 1956 suggested something very similar. Near the time of opposition, however, which occurred in March 1957, it became possible to distinguish a more definite section of the disturbed region. This was about 20° in length and both its ends were rotating in a period of about $9^h 55^m 21^s$, the longitude of its centre reaching 64° in System II by the end of April.

Even if this remaining feature should shortly disappear, it will be of the greatest interest to observe whether the next restoration of the S. component of the S.E.B. to its normal intensity will be accompanied by the reappearance in the S. Tropical Zone of a disturbed area that can be identified by its motion with the object that has just been described. But the interval must not be too long; after the disappearance of the S. Tropical Disturbance during 1926, 1927 and the early part of 1928 it was identified again, not by extrapolation from its previous motion but by the familiarity of its appearance, a criterion that would scarcely be valid in the case of the present object.

CHAPTER 16

THE DARK SOUTH TROPICAL STREAKS
OF 1941-42 AND 1946

Towards the end of October 1941 a remarkable dark object was seen to be forming in the S. Tropical Zone in the region immediately preceding the Red Spot. The first record of it was made by Ainslie on October 22 and it was noted independently by the author on October 23, three rotations of the planet later. It gradually drew out in the preceding direction into a long, heavily shaded streak until it was about 30° long and then, about the middle of November, its following end, which had remained stationary and had seemed to mark the p. end of the Red Spot Hollow while the streak was forming, also drew away from the Red Spot at a rate that allowed the total length of the streak to remain approximately constant until at any rate the end of March. The rates of decrease in System II longitude of the preceding and following ends were $10°1$ and $11°3$ respectively in thirty days, that of the f. end being reckoned from the middle of November only, when the length of the streak had become fully established. These convert to rotation periods of $9^h 55^m 27^s$ and $9^h 55^m 25^s$.

The development and subsequent behaviour of the Dark S. Tropical Streak may be followed by studying the series of sketches that are reproduced as Plate (IX)*; all of these, except the first which was made by the author, are due to Hargreaves, who prepared the drawings for the *B.A.A. Memoir*, Vol. 35, Part 3, in which they appeared as Plate V. The broken lines running from top to bottom of the plate represent the motions of the two ends of the Streak in longitude, the arrow-heads denoting longitude determinations to which there are no corresponding drawings.

It was suggested by Phillips that what we were witnessing was simply a revival of the South Tropical Disturbance; but perhaps 'rejuvenation' would have been a better word in view of the comparatively short length and rapid motion of the Streak. If the reader will compare some of the drawings of Plate (IX) with that of Plate VII, Fig. 1, which depicts the Disturbance as it appeared on 1903

---

* The original plates IX and X omitted for technical reasons from this edition.

August 12, he cannot fail to be impressed with the strong resemblance between the two objects. The rotation period of the Streak corresponded to that of the Disturbance in 1911, 1912 and 1913.

An interesting phenomenon connected with the Streak was the fading that began late in December of that part of the S. Temperate Belt which lay immediately to the south of it. There is a hint of this too in the drawing of 1903, though it was not a normal characteristic of the Disturbance.

Disappointment unfortunately awaited those who had been hoping that the once familiar S. Tropical Disturbance had taken on a new lease of life; for before the end of April 1942 something had begun to happen to the Streak, at any rate to its following end, that was causing it to grow noticeably shorter as the end of the apparition approached. When first seen again in the middle of August, it covered only about 13° and was still diminishing, the longitude determinations showing that during the interval the rate of progress of the p. end had remained sensibly constant and that the shrinking had taken place from the f. end of the Streak. By October 5 its length had been reduced to a mere 5° and it was last recorded by W. H. Haas on 1942 October 6. On October 11 H. M. Johnson, observing the region with the 18-inch refractor of the Flower Observatory, reported that no detail was to be seen in the South Tropical Zone.

Nothing further of particular interest was observed in the S. Tropical Zone until, on January 11, at the beginning of the apparition of 1946, Hargreaves detected a dark spot which, before he had ascertained its longitude, he thought might be part of the Red Spot. He saw it again on January 16, some distance away from the central meridian, and on January 26 the author was able to estimate its extension in longitude as about 12°; but after this an unprecedented spell of adverse weather conditions prevented further observations from being obtained until the last few days of March, when its length was 22°.

By this time the likeness of the object to the Dark S. Tropical Streak of 1941–42 was remarkable. It was still growing; and sketches of it, made by Hargreaves on April 9 and by the author on April 11, show a quite detailed resemblance to some of the representations on Plate (X). In particular, the S. Temperate Belt was again faint to the south of the Streak and the S. component of the S.E.B. curved southward to meet it at either end, helping to cause its extremities to appear concave towards the adjacent parts of the S. Tropical Zone and producing the effect that had enhanced the similarity between the Streak of 1941–42 and the old S. Tropical Disturbance.

Ultimately the length of the 1946 Streak surpassed that of 1941–42. By the end of the apparition in July it was covering nearly 40° of longitude and it seemed that this time it might really be going to establish itself as the old S. Tropical Disturbance; but when it was next seen, at the beginning of the apparition of 1947, it was shorter again and contracting rapidly, its length in the middle of April being only about 20°. It was last recorded on 1947 July 9, when it had diminished to roughly 6°. Thus the life of the second Dark S. Tropical Streak was longer than that of the first in the ratio of about three to two and at its maximum it was considerably more extended in longitude.

The rate of decrease of System II longitude of its p. end, while it was growing, was 14°5 in thirty days, that of its f. end 10°2; during the period of contraction these figures were found to have changed to 8°9 and 13°7 respectively, the inference being that this time both ends of the Streak were subject to the influence that was causing it to grow shorter. The mean rotation periods of the centre of the object were $9^h 55^m 24^s$ in 1946 and $9^h 55^m 25^s$ in 1947.

Although the Streak of 1941–42 seemed at its birth to be issuing from the p. end of the Red Spot Hollow, it is doubtful whether a similar origin can be attributed to the Streak of 1946. When first seen in January, the f. end of the latter preceded the p. end of the Red Spot by some 40°. It is perfectly possible, of course, that its career began in 1945 October, close to the Red Spot; but there is no evidence either for or against this hypothesis, as Jupiter was then too close to the Sun for observation.

CHAPTER 17

OUTBURSTS OF ACTIVITY IN THE SOUTH EQUATORIAL BELT

All the belts of Jupiter are subject to changes of intensity; any one of them may be conspicuous during one apparition and comparatively faint during the next, returning perhaps to normal after an interval of a few months or possibly a couple of apparitions.

These changes are generally so gradual that the observer may be unaware of their progress, unless he is keeping a systematic record of relative intensities. Sometimes the darkening may begin locally and spread gradually round the belt, during which process there may be quite definite preceding and following ends of the darker section.

The South Equatorial Belt seems to be unique, in that there have been several revivals from obscurity that have been almost cataclysmic in their origin and development. Five such outbursts have been recorded, the second and third of which will be discussed here in some detail.

The first to be recorded occurred during the apparition of 1919–20. During the latter half of the previous apparition the S. component of the S.E.B., including the Red Spot Hollow, had been gradually fading, as had also the S. Tropical Disturbance. When Jupiter became visible again after conjunction, none of these familiar objects could be seen, save that a few very faint fragments of the S.E.Bs. were occasionally glimpsed; meanwhile the whole space between the S.E.Bn. and the S.T.B. had become the brightest part of the disk, except that the Red Spot itself, which had previously been rather faint, had become a fairly conspicuous feature against the bright background. At the beginning of the second week of December 1919, however, a process of revival set in, starting with the outbreak of a number of dark humps at the S. edge of the N. component of the S.E.B. not far from System II longitude 230°. Subsequently a perfectly amazing amount of detail appeared, spreading over the whole width of the belt and becoming gradually extended round the whole 360° of longitude. On 1920 February 27 H. Thomson wrote: 'The S.E.B. is a most extraordinary spectacle. It consists largely of round dark dots and white spots.' Moreover, the changes of aspect were so rapid that it was almost impossible to identify the markings after so short an interval as a couple of days. Everything was turmoil and confusion; but during the latter part of March the Red Spot Hollow became visible again and at about the same time the two ends of the S. Tropical Disturbance reappeared in the S. Tropical Zone. It is unfortunate that more was not learnt about the individual motions of the features of this first dramatic upheaval; it seems likely that the clues were there but that they were overlooked because they were not detected in the early stages, as they were on the occasions of the second and third outbursts. An unpublished sketch by Thomson, for instance, dated 1920 January 1, shows a series of five dark spots spread out along the S. component of the S.E.B., the System II longitude of the first being 227°; these resemble so strikingly some of the objects that are about to be described in connection with the revival of 1928–29, that the obvious inference is that they must have been endowed with a similar rapidly retrograding drift in longitude.

As soon as Jupiter reappeared in the morning sky at the beginning of the apparition of 1926, it was noted that its aspect was very similar to that of the early months of 1919–20. The S. Tropical Disturbance, the Red Spot Hollow and nearly the whole of the S. component of the S.E.B. had disappeared; the S.E.Bn., though narrow, remained well defined and the Red Spot itself, as in 1919, gradually became more distinct and prominent. Two very short portions of the S.E.Bs. remained until well after opposition and became affectionately known to intimates as 'remnant' and 'fragment'; though separated by 80° of longitude, they were very similar in appearance and to distinguish between them, without previous knowledge of the longitude of the central meridian, was a feat to be achieved only by those who were on the most familiar terms with the planet. On Phillips' drawing of 1926 September 19, reproduced here as Plate X, Fig. 1, 'remnant' is shown approaching the p. limb while 'fragment' follows the shadow of Satellite III.

This time there was no immediate return of the missing features. Throughout the apparition of 1927–28 the appearance of the S. Tropical regions remained substantially unchanged. The S.E.Bs. was rarely seen as a very faint line; but the Red Spot was exceedingly conspicuous against a bright background that was generally continuous from the S.E.Bn. to the S.T.B. Close to the N. edge of the latter there ran in some longitudes a dusky line which, for a time, looped northwards in one place into the middle of the S. Tropical

Zone, earning from its appearance the name of the False Red Spot. This disappeared early in December 1927.

It was not until 1928 August 10 that the first sign of a revival of the familiar markings was detected in the form of a small dark spot in the latitude of the S.E.Bs. This appeared in System II longitude 128° and was connected to the S.E.Bn. by a slightly curving grey wisp. By the end of August there were at least four similar dark spots, situated in the same latitude and spread over about 80° of longitude following the position of the original object. It was not long before Phillips detected the extraordinary motion of these S.E.Bs. spots, whose System II longitude was increasing at the rate of 5° and more in twenty-four hours and which were running away from the longitude of the origin of the disturbance as if they had been ejected from it. The rotation period of the most rapidly retro-grading of these objects that was satisfactorily determined is represented by the unprecedentedly high value of $9^h\ 59^m\ 31^s$; but it is a reasonable inference from the chart that the period of the leading spot (spot of greatest longitude), which was not considered to have been well enough observed to render a reliable determination, may have appreciably exceeded ten hours!

Meanwhile, before the end of August, dark southward humps had begun to develop from the S. edge of the N. component of the belt, starting close to the same System II longitude as that of the original S.E.Bs. spot. Many of these became sharply pointed, their combined effect sometimes resembling the teeth of a saw; but changes were so rapid that it was not easy to keep track of the identity of individual features. The preceding end of this group, however, progressed rapidly and at first with an accelerated motion in the direction of decreasing System II longitude, showing a mean rotation period of a little more than $9^h\ 51^m$. There was, of course, nothing very remarkable about the latter motion, since the latitude concerned was adjacent to that of the S. Equatorial Current, which at that time had about the same period; but the general effect was that, relative to the point of origin of the whole disturbance, which, as judged by the longitudes where the rapidly retrograding S.E.Bs. spots continued to make their first appearances, showed slight acceleration with a mean rotation period of about $9^h\ 55\frac{1}{4}^m$, there were two streams of spots, one advancing along the S.E.Bn. at the rate of about 5° per day and the other retrograding along the S.E.Bs. at approximately the same speed. It must be emphasised, however, that during the first month of the upheaval the identification of individual S.E.Bn. spots was so difficult that, until after the middle of September, all that can be stated with certainty about the northern branch of the disturbance is that its p. end was advancing at the rate just given.

Inevitably, after the lapse of about six weeks from the date of the initial outburst, the two streams began to pass one another on the opposite side of the planet. Each group of spots retained its distinctive latitude; but the central regions of the S.E.B., which lay between them, then became a scene of turbulence and change such as had not been previously recorded, even in 1919–20. See Plate VIII, where various aspects of the disturbance are depicted, and especially Fig. 4.

We now return to the S.E.Bs. spots and their subsequent behaviour. It was clear that, if they survived long enough, they would encounter the Red Spot, which was then situated at about 317° in System II. Would they fade when they reached it, pass across it or perhaps be deflected around its northern edge?

As the spots proceeded on their course, the S. component of the belt seemed to be developing with them; and although the leader apparently faded before reaching the Red Spot, those following it were deflected northwards along the S. edge of the Hollow, which was beginning to reappear, becoming flattened out into short streaks during the process. Many of them survived the passage but all were lost after proceeding a few degrees beyond the f. end of the Hollow. The point of disappearance turned out to be close to the position where the p. end of the reviving S. Tropical Disturbance was ultimately identified, a fact that may seem to have considerable significance when considered in relation to the phenomena of the 'Circulating Current', which are detailed in the next chapter.

Soon after the passage of the first few spots around the S. edge of the Hollow, the Red Spot began to fade and the whole region to lose its characteristic features, after which some of the later spots, which also became elongated, were carried straight on across the area without suffering any northerly deflection; these also vanished after traversing a few further degrees of longitude.

By the end of the year the aspect of the S. Equatorial Belt was almost normal again and the Red Spot Hollow was a light oval surrounded by the duskiness of the S. Tropical Disturbance, the p. end of which had been observed from November 7 to December 4 preceding the Hollow, the long ninth conjunction having evidently begun during the first week in November.

A further noteworthy feature of the 1928–29 outburst was the extension of the influence of the more northerly current into the Equatorial Zone, where it not only affected the S. Equatorial Current but, for a short time (October 18–25), produced the unprecedented rotation period of $9^h 52\frac{1}{4}^m$ for at least two spots that lay almost in the middle of the Zone. Actually, from the middle of August until the end of the year, the S. Equatorial Current was divisible into two or three principal sections. Between System I longitudes 0° and 140° the spots at first had the very slow rotation period of about $9^h 52\frac{1}{2}^m$; this diminished gradually and progressively until, by the beginning of November, it had become almost exactly equal to that of System I. During September there was a group of spots from about 140° to 180° which, though apparently subject to sudden minor irregularities, preserved on the whole a far more uniform period of a few seconds longer than $9^h 51^m$ until most of them vanished at the end of December. Following these and starting between 180° and 210° towards the end of September, was a small group, whose motion became so retarded that the slowest of them showed a period as long as $9^h 53^m 37^s$ during the middle of October. It was at this time and in the same longitudes that the abnormally drifting objects near the middle of the Equatorial Zone, which have just been mentioned, were observed. But the most striking of all the S. Equatorial and S.E.Bn. features was a large and very dark elongated spot that became known as the 'blob' and is shown a little way past the central meridian on Phillips' beautiful drawing of 1928 September 11; see Plate V, Fig. 4. When first recorded on July 25, this was practically stationary in System I with its centre at 322°. About the middle of August, however, its motion became accelerated; but it slowed down again and was once more stationary at about 282° during the first week of October. It was just beginning to show a tendency to retrograde, when it was encountered by the last-mentioned group of spots with the very long rotation periods, with the result that it disappeared. It seems that these spots must have passed right over the 'blob'; for, after they had gone by, either the object itself or another dark and conspicuous spot appeared, whose line on the chart, showing retrograde motion, was an exact continuation of that of the original 'blob' just before it faded. This retrogression continued until the spot was last seen on January 21. Between the 'blob' and System I longitude 20° there were traces on the chart of one or two spots that may have had parallel motions. For the fullest details of these remarkable features reference should be made to the Twenty-seventh Report of the B.A.A. Jupiter Section and especially to the longitude chart which is reproduced there as Plate IX.

Discussion of one remarkable series of phenomena, which were associated with each of the first three recorded S.E.B. revivals, is postponed until the details of the eruption that took place during the apparition of 1942–43 have been described; we refer to the temporary display of white spots, that became so conspicuous at great distances from the central meridian that the effects of irradiation caused them to appear at times to be actually projecting from the limb or terminator of the planet.

Meanwhile it must not be overlooked that a somewhat similar disturbance may have occurred in the S.E.B. in the interval between the apparitions of 1937 and 1938 during the four or five months when Jupiter was too close to the Sun to be observed from the northern hemisphere. All through the apparitions of 1936 and 1937 the S. Tropical Zone had been bright, the S. Tropical Disturbance invisible and the S. component of the S.E.B. either invisible or extremely faint, while the Red Spot was exceedingly conspicuous. As soon, however, as the planet became observable again in May 1938, it was seen that the S.E.B. had returned to normal; the Red Spot Hollow was again presented as a complete oval, within which the Spot itself was invisible, and the p. end of the S. Tropical Disturbance had become a conspicuous feature, though the f. end, which must have been near conjunction with the Red Spot, was never definitely located until the next apparition. It will be seen, therefore, that there was just about time for a considerable S.E.B. disturbance to have broken out and died away again; but, if this were the case, it must either have started not later than early in January 1938 or have been on a scale inferior to its predecessors; for in May no obvious evidence remained of unusual activity in this region.

Towards the end of the apparition of 1939–40 the S. component of the belt began to fade again and it remained faint, though by no means invisible, until more than halfway through the apparition of 1942–43. During this period the S. Tropical Zone was again rather bright, though not outstandingly so, while the Red Spot never became really conspicuous.

From 1942 December 17 to 1943 February 6 a small grey projection with slowly increasing System II longitude had been observed from time to time at the S. edge of the S. component of the belt; this was recorded at about 17° by both Hargreaves and the author on February 4 and was seen again by both observers on February 6. On February 7, only two rotations of the planet later, this appeared to the author as a conspicuous dark spot, well to the S. of the S.E.Bs., and it was seen that, following it by about 2°, was a dark spot between the components of the belt which, if present on the previous night, must have been quite inconspicuous or it would not have been missed. By February 11 both these spots had become intensely dark; the more northerly one was then slightly preceding the other and was preceded itself by a white spot and then another dark one. Soon the whole structure of the S.E.B. in this region became complex, the development proceeding in a manner strikingly similar to that of 1928–29. Dark spots, which appeared to originate at about System II longitude 20°, moved with rapid retrograde motion along the S. component of the belt; but as these were subject to complicated changes among themselves and probably in some cases to subdivision, it was not possible to obtain as complete a record of their individual motions as was provided by their counterparts of 1928–29. The two leaders, however, one of which was observed from February 18 to April 18 and the other from February 18 to March 17, had rotation periods of $9^h 57^m 50^s$ and $9^h 57^m 36^s$; a third, which developed later in a greater longitude than either of these, had a somewhat longer period, if two subsequent identifications are to be trusted.

In 1928 the northern part of the disturbance moved along the N. component of the belt with a rotation period that was considerably longer than that of System I. On this occasion also its initial period of about $9^h 52^m 36^s$ was very long for the S.E.Bn.; but as the result, no doubt, of an event of major importance at the end of February, which is about to be described, the rotation of the leading part of the northern stream became suddenly accelerated to a period of $9^h 49^m 52^s$, which appears to be the record for this latitude.

This is what happened. Before the end of February the outburst seemed to have become well established along the lines of the 1928–29 eruption, but with the rather slower motions in System II of about $+3°$ per day for the S.E.Bs. spots and $-4\frac{1}{2}°$ for those on the S.E.Bn. Then, on February 27, just as the northern branch was reaching System II longitude 288°, there appeared suddenly in that longitude and at the N. edge of the S. component of the belt a brilliant white spot, which was first detected by Hargreaves. By March 2 another bright spot, which preceded it by 3° or less, had appeared at the S. edge of the N. component and following this was an intensely dark spot, from which a hard, dark streak ran somewhat obliquely to the S. component; in addition there was a small grey projection from the S. edge of the S. component—Plate V, Fig. 2. By March 4 the two bright spots had apparently coalesced to form one larger object well to the S. side of the N. component and it soon became evident that a second seat of eruption had established itself, from which another series of dark S.E.Bs. spots was issuing to retrograde along the S.

component, while the northern spots had become accelerated as mentioned above.

It has been noted that in 1943 the speed of retrogression of the original S.E.Bs. spots was decidedly inferior to that of their 1928–29 counterparts; but those emitted from the secondary source were moving far more rapidly, the longest rotation period recorded being 9^h 59^m 31^s, which is exactly equal to the longest period actually determined in 1928.

As before the two opposing streams began to pass by one another, this time after the lapse of about five weeks on March 15; but owing to the comparatively slow progress of the southern spots the meeting place was far from being diametrically opposite to the site of the original outburst but at about 110° in System II.

In 1943 there was far less confusion and rapid change among the spots and markings that appeared in the middle of the belt than there had been on either of the previous occasions. Indeed, the light space between the components in the longitudes that followed the scene of the original outburst remained practically undisturbed right up to the Red Spot, which was situated at 172° in System II, even while the northern spots were flowing past it on their way to complete the encirclement of the planet. In the rest of the region the outstanding feature was a pair of white spots that became conspicuous about the end of February and appear to have originated a little earlier near the seat of the initial outburst. At first the leading spot lay definitely to the S. of the N. component, the following one being situated in the S. half of the space between the components; but, as their System II longitudes rapidly diminished, both moved northwards, until the first lay right on the N. component and the second only a little to the S. of it. It was almost certainly the latter that gave rise to the effect of irradiation at the terminator, which is shortly to be described. Their mean rotation period was 9^h 52^m 28^s.

There remains for mention the behaviour of the retrograding S.E.Bs. spots of 1943 when they reached the longitude of the Red Spot. By March 28 the leading (f.) spot, which had become drawn out into quite a long streak, had almost reached the p. end of the Red Spot, though the preceding half of the latter had faded, leaving only its following half visible; but the weather in the British Isles, which had been astonishingly favourable ever since the beginning of the outburst, chose this critical time to thwart the observers in their efforts to ascertain whether the phenomena of 1928 would be repeated. For four consecutive nights no observations were recorded and, although drawings made on April 2, 3 and 4 show rather violent

166

changes to have been taking place in the region, nothing that could be recognised as the streak was detected. On April 6, however, a definite streak, with its leading end in exactly the right longitude, was detected in the S.E.B. 'space' to the north of the Red Spot, whose whole outline was again visible. By April 11, when the area was next seen, a streak in the right longitude had completely formed on the S. component immediately following the Red Spot, while to the north of the latter a second one was visible. The former continued its retrogression, its track on the longitude chart being consistent with the presumption that it was identical with the original leader, while on April 13 another dark streak seemed to have formed ahead of it. On several other occasions dark spots were seen to the north of the Red Spot but it was impossible either to confirm or deny their correspondence with the original members of the series. An excellent idea of the whole course of events may be obtained from Plate X, which was prepared by Hargreaves from a series of strip-sketches by himself and the author for the 33rd Report of the B.A.A. Jupiter Section (*Memoirs*, Vol. 35, Part 4) where it appears as Plate V. The cross indicates the outbreak of the secondary disturbance.

The Irradiating Spots

Associated with each of the outbursts already described was the remarkable phenomenon of white spots that were so little subject to the 'limb darkening', which affects the whole of the periphery of the planet and is emphasised every time a satellite begins or ends transit, that the contrast of their brightness with that of their background caused them to appear by irradiation to project beyond the limb or terminator.

We shall deal first with the Spots of 1928 and 1943, because the circumstances were similar in all essentials. Each of these upheavals had one such spot associated with it; on both occasions the spot was situated just within the N. edge of the S.E.B. and on neither was any irradiation observed until the outburst of activity in the belt was fully developed, in 1928 about ten weeks and in 1943 about five weeks after the beginning of the eruption.

In 1928, although there was only one irradiating spot, its apparent projection was seen at both the preceding and the following limbs and, as the planet was practically at opposition at the time, the effect of the terminator was negligible. It was first observed on October 24, when going off the disk at the p. limb; on the following night and again three nights later it was seen at both limbs. On October 30 it was seen irradiating for the last time; for although it appeared bright at the limb on three other occasions up to November 8, no further projection was noted.

The times at which irradiation began and ended indicate that it was never observed when the source lay within 15° of the true limb. The duration of the phenomenon averaged about 25 minutes, during which interval Jupiter rotates through some 15°. Since this implies that by the time irradiation ceased at ingress the original source was 30° from the limb, it can be inferred that the white cloud that produced the effect must have had considerable extension in longitude; for at 15° a point on the surface of the planet would have appeared to lie only about 0·8 seconds of arc within the limb, while at 30° the same point would have increased this apparent distance to three seconds, which is surely too far for it to have given rise to the spurious effect of projection. The converse holds, of course, for egress.

In a paper dealing mainly with these phenomena (*M.N.*, 89, 708, 1929) Hargreaves attempted to establish the identity of a very bright spot at the following end of a bright area in the S.E.B., which was repeatedly seen on the central meridian between September 30 and December 4, with the object that gave rise to the irradiation; its longitude, however, proved to be about 5° too small, unless it was assumed that the phases at the p. and f. limbs where irradiation occurred were unsymmetrically placed with regard to the central meridian. In the same number of *Monthly Notices* (p. 703) there is a paper by W. H. Wright, in which photographs of Jupiter, taken in the light of different wavelengths during October and November 1928, are reproduced; some of these, notably one dated October 28, show in ultra-violet light a vast bright cloud, greatly extended in longitude, the brightest part of which, near its f. end, is in the right position to have been the cause of the irradiation and is more than large enough to have included the bright spot observed visually on the C.M. Doubtless the three phenomena were interrelated. Objects that appear bright in the ultra-violet almost certainly have relatively high elevations in a planetary atmosphere, an inference that would account for their being less affected by limb darkening than the normal cloud layer lying beneath them; but the cloud that was photographed was so large that only a relatively small portion of it can have been concerned in producing the irradiation.

In 1943, although the effect of irradiation was first observed only five weeks after the beginning of the disturbance, opposition had occurred more than two months previously, with the result that, during the whole period over which the phenomenon was recorded, the terminator lay some 10° within the true f. limb. At such a time the p. limb would have been somewhat brighter than either limb at opposition in 1928 and it is for this reason, no doubt, that all the records of apparent projection in 1943 relate to the terminator and that no unusual effect was ever noted at the limb, although observers were keenly on the watch at the times when it was to be expected.

As in 1928, there was only one spot or small area that gave rise to the appearance of irradiation, which was first detected on March 13. Subsequent observations were obtained on March 16 and 27 and on April 3 and 4. The latitude of the projecting object was estimated on the first two occasions to be that of the space between the components of the S.E.B. and subsequently that of the N. component. This is consistent with its having been associated with the following member of the pair of bright spots, mentioned above, that were repeatedly recorded on the central meridian; for during the period covered by the observations the southern latitude of this spot was gradually decreasing. Unfortunately no ultra-violet photographs are available for March and April 1943. Meridian passage of the bright spot occurred on the average 66° after irradiation with no residual greater than 4°, which places the irradiating phase as about 14° within the terminator. Shortly after the beginning of April the two bright spots faded; but a revival took place about April 18, which was followed by a remarkable observation by Hargreaves, who on April 22 detected irradiation at the terminator in full daylight. This achievement was rendered possible by the employment of a Polaroid filter, which under the right conditions can be made to reduce the light of the sky very appreciably and was particularly effective at this time, since Jupiter was about 90° from the Sun where the polarisation of sky light is a maximum. On this last occasion the bright spot would have been some 18° within the terminator.

We return now to the apparition of 1918–19, when at least three separate objects were seen to give rise to the effect of projection. The analogy with the phenomena of 1928–29 and 1942–43 is far from being complete; but it is difficult to believe that the association of these spots with the behaviour of the S.E.B. was entirely fortuitous. It will be remembered that the S.E.B. revival of 1919–20 was preceded by only a short-lived fading of the S. Tropical markings, which set in during February 1919. It was actually just before and during the initial stages of the *fading* that the irradiating spots were observed, none being recorded during the revival; moreover the spots were situated on the Equatorial Zone, one of them lying even a little to the north of the equator. None was identified on the central meridian;

but as all of them were seen at both limbs, it was possible to estimate their approximate longitudes and to determine the projecting phase as being some 70° to 80° from the C.M. Opposition occurred on 1919 January 9.

The first spot was observed five times from 1918 December 29 to 1919 January 17, the third four times from January 26 to February 12; both had mean rotation periods of the order of 9^h 53^m. The second, which was the most northerly, was seen six times at the terminator and four times at the limb from January 16 to 26; its mean rotation period was close to that of System I, in which system its longitude seemed to vary capriciously between about 260° and 285°, which suggests that the irradiation may have been attributable at various times to different portions of a fairly extensive, high-level white cloud.

Subsequent Revivals of the S. Equatorial Belt

Between 1943 and the time of writing there have been two further outbursts of a similar nature in the S.E.B. The former of these took place during the apparition of 1949, when Jupiter lay in high southern declination, and it is unfortunate that, for our knowledge of what occurred, we have to rely on the record of a single observer. In a typical temperate climate it is almost impossible for a single unaided individual to deal adequately with phenomena such as these. The relatively full and complete accounts of the events of 1928 and 1943 that we have been able to present are due to the combined efforts of a little group of observers, who were working at places sufficiently remote from one another to ensure that overcast skies or bad definition in one locality did not always result in gaps in the record. If any one of the observers of these earlier years had been working alone, our knowledge of the character of these outbursts might have been decidedly meagre.

Students of Jupiter have every reason, therefore, to be grateful to Mr. R. A. McIntosh, of Auckland, N.Z., for his single-handed effort, without which we should have presumably remained in ignorance of even the fact that any remarkable phenomena had been displayed. Equipped with a 14-inch reflector, he made a large number of invaluable observations and has been able to give us a very fair picture of the main features of the disturbance. With Jupiter in nearly its maximum southerly declination, it is a misfortune that there were no other observers anywhere in the southern hemisphere, who were both willing and able to join forces with him in the way that the northern observers had been co-operating with one another.

170

Under the circumstances it is not surprising that our knowledge of the events attending the revival of 1949 is less detailed than it is for 1928 and 1943.

The outburst of 1949 began either with a dark spot, seen on the S. component of the S.E.B. in System II longitude 157° on July 12, or with a small bright spot, first recorded on July 19, which formed, also on the S. component, at 163° after the other had disappeared. Apparently this bright spot expanded to fill the whole of the space between the components, while its System II longitude was decreasing in accordance with a rotation period of $9^h 54^m 45^s$. As soon as it had moved far enough, a similar white spot formed behind it between the components and in about the original longitude, to be followed by another and another, each in turn setting off in pursuit of the leader. At least twenty such spots were seen and the rotation periods of fifteen of them, ranging from $9^h 54^m 51^s$ to $9^h 52^m 29^s$, were determined. Meanwhile the space between the components around these bright spots had been darkening and the rotation period of the preceding end of the whole disturbance was found to be $9^h 53^m 50^s$.

As on previous occasions there were some dark spots moving along the S. component of the belt. McIntosh gave $9^h 57^m 25^s$ as the mean rotation period derived from six of these objects; but his published chart (*B.A.A.J.*, Vol. 60, 247, Plate VI) leaves one in some doubt as to the correctness of his identifications. It appears to the author that one such spot, of which there seem to have been three reliable observations, had a period close to $9^h 58^m$.

On the whole it does not seem permissible to regard the outburst of 1949 as a close parallel to that of either 1928–29 or 1942–43 for the chief reason that the most conspicuous feature of 1949 was the long string of discrete and identifiable bright spots that ran along in the p. direction between the components of the S.E.B., producing a general appearance quite unlike the aspect of the region associated with any of the previous upheavals.

The following summary of the main features of the disturbance of 1952–53 has been compiled after reference to unpublished* records and must therefore be regarded as provisional. It began with the appearance on 1952 October 22 of a dark spot, situated between the components of the S.E.B. in System II longitude 204°. The preceding end of this disturbance advanced along the N. component with a rotation period not very different from that of System I, while many dark spots appeared on the S. component in the longitudes

---

* A discussion of these observations is now to be found in the 39th Report of the B.A.A. Jupiter Section, *B.A.A.J.* 64, 281 and 379, 1954.

following that of the original spot. The observations were numerous but just lack the continuity necessary for providing unassailable identifications of the S.E.Bs. spots. If this eruption had stood alone, the longitude chart would have presented a problem in identification that would probably have remained unsolved; but on the assumption, in the light of previous experience, that these spots were retrograding along the S. component, it is possible to follow several of them with some measure of confidence and the author has determined approximate rotation periods for four of them as follows:

$$
\begin{array}{llll}
(1) & 9^h & 58^m & 42^s \\
(2) & 9 & 58 & 17 \\
(3) & 9 & 57 & 39 \\
(4) & 9 & 57 & 3
\end{array}
$$

The first spot led the retrogression and the others followed in order. All appear to have succeeded in passing the Red Spot, whose longitude at the beginning of November was 270°, and a drawing dated November 16 by E. J. Reese, who was responsible for a large number of the observations, shows one such spot apparently 'descending' the p. shoulder of the Hollow. In addition to these, several dark S.E.Bs. spots were recorded in the longitudes preceding that of the original outburst; but there is no convincing evidence that a secondary source, similar to that of 1943, was responsible for them. During the later stages of development a large number of white spots were observed in the space between the components but the difficulties in assigning identifications appear to be insuperable.

All things considered, it seems likely that the 1952–53 revival of the S.E.B. was similar in most respects to that of 1942–43.

In the theoretical portion of this book will be found a discussion of an interesting attempt by Reese to identify the original longitudes of all these upheavals with a single definite locality on the solid surface of the planet.

CHAPTER 18

THE CIRCULATING CURRENT

The events described in this chapter present to the student of planetary atmospheres a problem that is probably the most baffling and intriguing of all the many mysteries that have yet to be solved in the realm of Jovian meteorology.

It is best to treat them chronologically and we therefore return to the latter part of the apparition of 1919–20, when the first of the great revivals of the S. Equatorial Belt was in progress. As was stated in the last chapter, the identification of spots during the first few weeks of the outburst was rendered exceedingly difficult by the general confusion of detail and the rapidity of the changes that were taking place; but during February 1920 two conspicuous dark spots appeared, which at first were threaded upon the S. component of the S.E.B. but later, though still practically in contact with this component, lay slightly to the south of it. These spots were separated by about 60° of longitude and had rapid retrograde motions. The leader (spot of greater longitude), which has been designated A, had a rotation period of $9^h 57^m 59^s$, which at the time was the longest on record; the period of the other, known as B, was $9^h 57^m 52^s$, so there was little change in their separation in longitude. Both were moving rapidly towards the p. end of the lately revived S. Tropical Disturbance which, relative to System II, was advancing slowly to meet them. On March 20 the spot A was recorded very close indeed to the p. end of the Disturbance, after which it suddenly vanished and was never seen on the following side of it. The spot B continued to approach the p. end but, presumably on account of cloudy weather, was last seen on March 30, when the distance it had still to travel was about 25°; conjunction with the p. end would have taken place about April 6. Like A, this spot was never seen on the following side of the p. end.

On March 23, only three* days after A was last seen at the S.

* Surely then either the spot a has been placed too far to the left on Smith's drawing of March 23—see Plate III, Fig. 7—or the date is wrong. There is inconsistency in the account given in the *B.A.A. Memoirs*, where in another place the date when a was first recorded is given as March 26.

edge of the S.E.B., a dark spot, smaller than A but by no means inconspicuous, was seen for the first time in the extreme south of the S. Tropical Zone, almost in contact with the N. edge of the S.T.B. and preceding by a few degrees the p. end of the Disturbance. This spot, which is referred to as *a*, had a rapid motion in the preceding direction and a rotation period of $9^h 53^m 29^s$; in other words it was running away from the p. end of the Disturbance at almost the same speed as that at which A had been and B still was approaching it. On April 13, a week after B should have been in conjunction with the p. end, a second spot, closely resembling *a*, was recorded at the N. edge of the S.T.B. in longitude about 20° less than that of the p. end. The rotation period of this spot, known as *b*, was $9^h 53^m 20^s$; so it too was running away from the S. Tropical Disturbance at almost the same rate as *a*, upon which it was actually gaining very slowly and which at the time preceded it by about 55°. These two spots continued to skirt the N. edge of the S.T.B. until they were lost to view near the end of the month, *a* being last seen at about 35° in System II on April 25 and *b* at about 75° on April 27.

Figs. 5 to 9 of Plate III show the aspects of these four objects at various stages in their history. In Figs. 6 to 9 the left-hand edge of the dusky marking in the S. Tropical Zone is the p. end of the S. Tropical Disturbance and it should be noted that in Fig. 8 the spots B and *a* are shown simultaneously. Fig. 6 on p. 171 is the chart of their motions in longitude, together with that of the p. end of the S. Tropical Disturbance; except for the dotted extensions, it is similar to that given in the Twenty-second Report of the B.A.A. Jupiter Section, which deals with the apparition of 1919–20 but which was not published until 1926.

It was while studying this diagram in the 1919–20 Jupiter Report that the author was struck by the fact that, when the tracks of the spots were produced, as shown by the dotted lines, the intersection of A with *a* and B with *b* fell extremely close to the track of the p. end of the S. Tropical Disturbance. This suggested that the latter feature might not only have provided a barrier to further retrograde motion by A and B but might have acted in some inexplicable manner as a reflector, turning them back with equal but reversed velocities after having somehow transferred them in the process from the N. to the S. side of the S. Tropical Zone, in which case *a* was simply A upon its return journey and *b* was B. In view of the concave form exhibited by the p. end of the Disturbance a good, but of course purely pictorial analogue of the idea is the progress of two balls, projected along one side of a bagatelle table, which would run round the semicircular end and back along the opposite side.

Considered as an elementary exercise in reading a graph, everything tallied beautifully (*B.A.A.J.*, 37, 62, 1926); but as an interpretation of the actual observations the notion seemed, to its

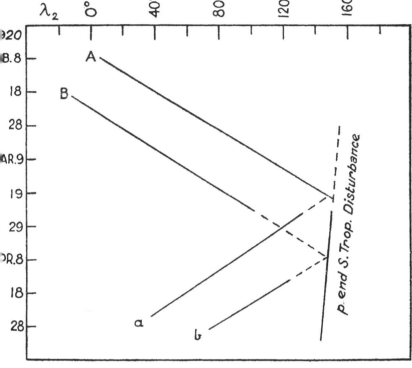

FIGURE 6.—The motion in longitude of the "Circulating Spots" of 1920.

originator at any rate, little short of fantastic. However, as the President of the B.A.A., Dr. W. H. Steavenson, observed when commenting on the author's paper, if it was not after all the right key, it was certainly a marvel that it should open the lock so readily. In any case it provided an incentive to observers to be on the watch for similar phenomena and we shall shortly see how their vigilance was ultimately rewarded.

Naturally, when the rapidly retrograding S.E.Bs. spots of the 1928–29 eruption made their appearance, the possibility that they might be analogous to the spots A and B of 1920 was not overlooked.

Unfortunately the recognition of the p. end of the S. Tropical Disturbance was delayed, on account of its faintness and indefinite character, until late in the apparition, when the outbreak of S.E.Bs. spots had subsided. It was then found that it had actually been observed several times during the period of their activity and, as was mentioned in the last chapter, that it had occupied the longitude at which their arrival had been accompanied by their extinction. We shall refer later to a further aspect of its probable relation to these spots.

The succeeding apparition produced no interesting spots at the S. edge of the S.E.B.; but towards the end of 1930–31, in March and April 1931, three or four dark humps appeared at the S.E.Bs. that were endowed with very rapid retrograde motion, rotation periods of $9^h 59^m 14^s$ and $9^h 59^m 6^s$ being determined for two of them. They duly reached the p. end of the S. Tropical Disturbance and never before can the N. edge of the S.T.B. have been scrutinised so minutely. But to the profound disappointment of all concerned nothing whatever happened! It seems certain that the spots did not pass beyond the p. end; but no recognisable sign was seen of a returning spot at the S.T.Bn.

Of the apparition of 1931–32, however, there is a different story to tell. Towards the end of February 1932 three more small, dark and rapidly retrograding humps were identified at the S. edge of the S.E.B. in the region preceding the p. end of the Disturbance. They duly reached the p. end and disappeared, one after another, and once more the attention of observers was focused upon the N. edge of the S.T.B. Writing more than twenty years after the event, it is still impossible to record without recapturing some of the thrill experienced by observers at the time how, early in March, three dusky objects were detected, lying just to the N. of the S.T.B. and moving rapidly in the preceding direction away from the p. end of the Disturbance. They took the form of short grey streaks and, although they were quite unlike the S.E.Bs. humps in appearance, their tracks on the longitude chart, when produced backwards, intersected those of the S.E.Bs. humps 3° or 4° on the following side of the p. end of the Disturbance.

Meanwhile a number of new humps had appeared at the S.E.Bs. and, with one exception only, all were carried southwards and 'reflected' by the Disturbance. Figure 7 reproduces the chart, given in the Twenty-ninth Report of the B.A.A. Jupiter Section, of the longitudes and motions of the S.E.Bs. and S.T.Bn. spots from January to May 1932, the former being plotted as dots and the latter as crosses. It should be particularly noted that the irregularities in the spacing of the first five S.E.Bs. humps are faithfully reproduced in the spacing of the first five S.T.Bn. streaks (the

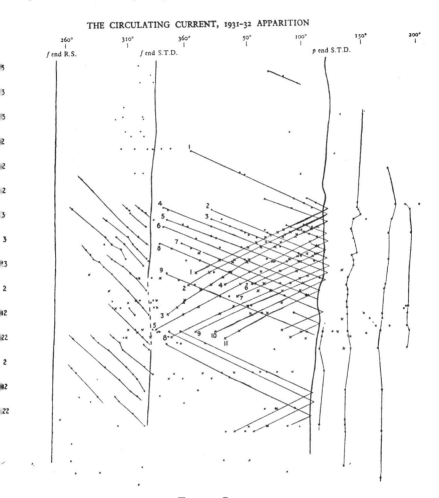

FIGURE 7

numbering of the spots on the chart refers to the order in which they were given in the Tables of Rotation Periods in the Report and are not for the purpose of identification; e.g. No. 1 at the S.E.Bs. becomes No. 2 upon its reappearance at the S.T.Bn. and No. 6 becomes No. 11). The short distances to the right of the track of the Disturbance, at which the intersections take place on the chart, are to be interpreted as representing the times occupied by the spots in traversing the S. Tropical Zone from north to south.

In view of the strong evidence which by this time had accrued that a single drift in Jupiter's atmosphere was sweeping the objects along the S. edge of the S.E.B., then around the p. end of the S. Tropical Disturbance and finally back in the opposite direction close to the N. edge of the S.T.B., it seemed appropriate to name this stream 'The Circulating Current'. Strictly interpreted, this name would imply that when, at the end of their return journey, the S.T.Bn. spots reached the f. end of the Disturbance, they were carried back to the S.E.Bs. to begin a second 'circuit'. Actually no such event was ever definitely recorded; but the appellation, though possibly premature, has undoubtedly come to stay. Concerning the relations of the spots with the f. end more will be recounted shortly.

During the apparition of 1932–33 there was a great profusion of spots in both branches of the current, the intersections of their tracks on the chart occurring regularly on or just behind the track of the p. end of the Disturbance, whose System II longitude was subject to some capricious variations of a few degrees in extent. The appearance of the p. end also underwent frequent changes; during March and April 1933 it was usually represented by the p. end of a conspicuous, dark, elongated and often detached streak, which lay only a little to the N. of the middle of the S. Tropical Zone. Though none of the 'circulating' spots was seen actually in mid-transit from one branch of the current to the other, some idea of the initial and final stages of the transfer may be obtained from a careful scrutiny of Plate IX. This shows a number of drawings of the region close to the p. end of the Disturbance, beautifully executed by Phillips and Hargreaves during a spell of exceedingly favourable observing conditions in March and April. The arrow in each case indicates the p. end of the Disturbance (note the dark streak just mentioned) and particular attention is invited to the behaviour of the objects labelled a, b and c. In Figs. 1 and 2 a is seen as a hump at the S. edge of the S.E.B., some distance preceding the p. end of the Disturbance. In Fig. 3 a has arrived at the p. end and is followed by another hump b. Two days later (Fig. 4) a has appeared as a dark streak on the S. side of the S. Tropical Zone, while b has approached much nearer to the p. end of the Disturbance. In Fig. 5 a has travelled some distance to the left, while b seems to have become tapered to a sharp point, preparatory to crossing the zone. After two more days (Figs. 6 and 7) a is almost out of the picture and yet another hump, *c*, at the S.E.Bs. has come into the region from the left; *b*, having almost completed its transit of the zone, is seen at an inclined angle on a grey wisp that seems to lead out of the p. end of the Disturbance. This record of *b*, as delineated in Fig. 6 and confirmed by Fig. 7, represents the nearest approach, known to the author, to an observation of one of these spots in actual transit of the zone. By April 3 (Fig. 8) *a* has passed out of range to the left, *b* has completed its crossing of the zone and now lies parallel to the direction of the belts and *c* is nearing the p. end. One rotation of the planet later (Fig. 9) *b* is farther to the left and *c*, almost at the p. end, appears to be fading. On the following night (Fig. 10) *b* is still seen but there is no certain indication of *c*. By April 8 (Fig. 11) *b* has passed out of view and *c* appears as a slightly inclined darkening of the curved grey wisp issuing from the p. end of the Disturbance.

The same activity at both S.E.Bs. and S.T.Bn. persisted throughout the apparition of 1934 and was in full vigour as late as the middle of June of that year. The fading of the whole of this remarkable phenomenon must therefore have been rather sudden; for at the beginning of the next apparition in January 1935 not a trace was to be seen of a spot in either branch of the current.

In 1938, however, a rather remarkable thing happened. It will be remembered that, when the planet was first observed after conjunction with the Sun, the S. Tropical markings were seen to have returned again to visibility. The p. end of the Disturbance was well observed during the early months of the apparition, though the f. end was never definitely located. Naturally a sharp look-out was kept for spots in the Circulating Current but none was seen at the S.E.Bs. At the S.T.Bn., however, five dusky spots were followed and were found to be moving at the rate appropriate to the southern branch of the current, all having rotation periods close to the mean of $9^h \, 53^m \, 13^s$. The apparition of 1938 seems, therefore, to have been complementary to that of 1930–31, when spots were seen at the S.E.Bs. but not at the S.T.Bn.

We shall now consider the relation of these spots to the following end of the S. Tropical Disturbance. In 1931–32 the f. end was faint and difficult to observe; nevertheless its position was frequently determined and its track is shown on the chart of the Circulating Current—Figure 7. Now it will be seen, on referring again to this chart, that in the longitudes between the Red Spot and the f. end of the Disturbance many S.E.Bs. humps were recorded and that they too were endowed with rapid retrograde motions which, however, were slower by nearly two minutes in rotation period than those of the spots in the region following them, between the two ends of the Disturbance. Phillips considered at the time that these humps passed across the longitude of the f. end, being subjected in the process to a further lengthening of rotation period; but neither Hargreaves (*B.A.A.J.*, 49, 334, 1939) nor the author could ever subscribe to this view, maintaining that the more slowly moving humps vanished upon reaching the f. end of the Disturbance and that the origin of the more rapid ones was independent of them. It is true, of course, that some of the S.E.Bs. tracks on the chart can be continued right through the f. end; but is this more than fortuitous? The reader should study the chart for himself and consider the evidence, noting especially the misfit of spot No. 8 among others of the S.E.Bs. objects in the true current. If Phillips' contention were correct, it would imply a sad distortion of the symmetry of the 'Circulating Current'.

It will be seen that spots at the N. edge of the S.T.B., as shown by crosses on the chart, are also to be found between the Red Spot and the f. end of the Disturbance; but these cannot in general be connected with the returning S.T.Bn. streaks, the majority of which faded while still some distance from the f. end. The returning streak No. 5, however, was seen very close to the f. end and cannot be identified beyond it. Moreover, it may easily be considered to have been carried again to the S.E.Bs. and to be represented there by the lowest of the tracks on the chart; but again this is not evidence.

During the apparition of 1932–33 the exact position of the f. end of the Disturbance was rather uncertain, as it was recorded as such once only, early in the apparition. Its longitude at opposition was believed to coincide with that of the following end at 296° in System II of a dark portion of the S. component of the S.E.B. At this time there lay, just to the S. of the middle of the S. Tropical Zone, three large, dusky, elongated masses, whose centres occupied the longitudes 285°, 306° and 333°. The last of these, which was of course the first to be encountered by the S.T.Bn. spots, provided some sort of barrier to their progress; for none was seen on its preceding side. Thus the returning streaks faded some 40° before they reached the probable position of the f. end of the Disturbance. During this apparition and again in 1934 S.E.Bs. humps continued to be recorded between the Red Spot and the f. end of the Disturbance. In 1934, however, the speed of their retrograde motion was greatly reduced and with it the probability of their being identifiable with any of the true members of the Circulating Current.

In 1934 the f. end of the Disturbance was once again readily observable; but determinations of its longitude showed rather large and capricious variations. Several of the S.T.Bn. spots were seen to approach it quite closely and it is possible to get a number of intersections on the chart that are consistent with their having returned once more to the S.E.Bs.; but the rather erratic record of the longitude of the f. end would render attempts at identifications somewhat precarious, even if it were permissible to rely on intersections alone.

Frequently during 1934 the grey wisp, which appears in some of the sketches of Plate IX to be issuing from the p. end of the Disturbance and is well shown on Plate V, Fig. 3 and Plate VII, Fig. 7, continued its course like a thin N. component of the S.T.B. until it reached the f. end of the Disturbance; into this it swept in a curve, more or less symmetrical to that at the p. end. This feature has already been referred to as the 'smoke stack' in the section devoted to the S. Temperate Belt, as has also the fact that the returning S.T.Bn. spots travelled along it. One observation, made by the author on 1934 May 27 and confirmed independently by Phillips, may be of significance regarding the possible return of the S.T.Bn. spots to the S.E.Bs.; on that occasion he detected one of the former very close to the f. end of the Disturbance and referred to it in his record as 'coming down the "smoke stack"'.

We now glance back for a moment at the two apparitions when the southern branch of the Current might have been expected to appear but failed to be recognised. In 1930–31 the watch was so keen that no definite S.T.Bn. spots are likely to have been missed; but it is surely significant that towards the end of March 1931 the 'smoke stack', issuing from the p. end of the Disturbance, became quite an easy object, although it does not appear to have been recorded previously during that apparition. On March 26 Hargreaves actually noted that it was broken and discontinuous but he was unable to obtain the longitude of any fiducial reference point.

Going back still further to 1928–29, we remember that many of the retrograding S.E.Bs. spots managed to survive their passage through the longitudes occupied by the Red Spot, the earlier ones by following the curve of the Hollow, the later by proceeding straight across the region, and that, after travelling a few degrees beyond its f. end, they vanished at a longitude that was later recognised as having been that of the p. end of the Disturbance. Now many detailed drawings were made of this region (see Plate VIII of the Twenty-seventh Report of the B.A.A. Jupiter Section) and a careful examination of these reveals that during October 1928 there were two or three instances of dusky spots having been recorded, just to the N. of the S.T.B. in the short space between the f. end of the Red Spot and the p. end of the Disturbance. It seems highly probable, therefore, as was first perceived by Hargreaves (*B.A.A.J.*, 49, 334, 1939), that the spots did get round the p. end to begin their return journey but that after two or three days they were obliterated, not by the Disturbance but by the Red Spot, whose southern edge was almost in contact with the S.T.B. One of the best examples of a possible returning S.T.Bn. spot in 1928 is illustrated here in Fig. 10 of Plate III, where an arrow marks the longitude of the p. end of the Disturbance and the letters R.S. indicate the position of the Red Spot, which has become almost unrecognisable amid the general turmoil in the region.

We may now ask ourselves what really reliable evidence we have that each returning S.T.Bn. spot was physically connected with its counterpart at the S.E.Bs. The changes engendered by their encounters with the p. end of the Disturbance were too great for the spots to be recognisable by characteristics such as shape or even size. We have to rely, therefore, on discontinuities and irregularities in their distribution or on the isolation of individual objects. From 1932 to 1934 the only discontinuity was the beginning of the whole series. This appeared to be well marked in both branches of the Current, though there is evidence that one or two isolated spots at the S.E.Bs., of which no returning counterparts were recorded, had appeared a little earlier in the apparition of 1931–32. There is also the strong confirmation supplied by the irregular spacing of the first five spots of the series. In the author's opinion, however, the really overwhelming evidence that the so-called 'Circulating Current' was indeed a physical reality lies in the magnificent isolation of each of the two (or, if the reader prefer it, four) spots of 1920; there was never the least possibility of mistaken identity. It may be added, perhaps, that none of those who observed these phenomena systematically had the slightest doubt but that the course of events was essentially as has been described, even though their explanation at present baffles the imagination.

After the last of the 'circulating' spots had vanished at the end of 1934, students of Jupiter were naturally wondering whether they would be observed again and, if so, whether their next appearance would throw more light upon their relation to the following end of the S. Tropical Disturbance. The five S.T.Bn. spots of 1938 revealed nothing; and since the total disappearance of both ends of the Disturbance in 1940 the chances of a recurrence of the phenomena seem to have dwindled almost to zero. One can only hope that in due course the S. Tropical Zone may give birth to a new Disturbance,* without which it is difficult to imagine a repetition of the phenomenon of 'circulation'; as has already been pointed out, examination of early drawings of the zone imply that such a renascence may not be utterly improbable.

Between 1940 and the time of writing, we have, of course, had the rapidly retrograding spots of three more revivals of the S.E.B. but with no sign of the returning current. There is also fairly strong evidence that in 1947 three S.E.Bs. spots had a mean rotation period of $9^h 58^m 34^s$ and that in 1941–42 there were four S.T.Bn. spots with a mean period of $9^h 52^m 30^s$; but that is all.

Of the three accompanying Tables, the first and second give the mean rotation periods of spots in the two branches of the Circulating Current during the apparitions when either branch is believed to have been displayed; in the third are found the periods of other retrograding spots at the S. edge of the S.E.B.

Northern Branch of the Circulating Current
Spots at the S. edge of the S. Equatorial Belt
(be ieved to be true members of the current)

Apparition	Rotation Period	No. of Spots
1919–20	9^h 57^m 55^s	2
1928–29	58 34	18
1930–31	59 10	2
1931–32	58 58	9
1932–33	58 48	32
1934	9 58 53	9
	Mean $9^h 58^m 43^s$	

Southern Branch of the Circulating Current
Spots at the N. edge of the S. Temperate Belt
(believed to be true members of the current)

Apparition	Rotation Period	No. of Spots
1919–20	9^h 53^m 24^s	2
1931–32	52 37	11
1932–33	53 0	14
1934	52 54	9
1938	9 53 13	5
	Mean $9^h 53^m 2^s$	

* See Addendum to Chapter 15.

Other Spots with long rotation periods at the S. edge of the S. Equatorial Belt

Apparition	Rotation Period	No. of Spots
1931–32*	9^h 57^m 4^s	16
1932–33*	57 14	15
1942–43†	58 34	5
1947	9 58 34	3

* Spots in the longitudes between the Red Spot and the f. end of the S. Tropical Disturbance.

† Spots of the great revival of the S. Equatorial Belt.

CHAPTER 19

THE OSCILLATING SPOTS OF 1940-41 AND 1941-42

During the apparition of 1940–41 a small but at times very dark spot, about the size of a satellite disk, made its appearance in the southern part of the S. Tropical Zone, fairly close to the N. edge of the S. Temperate Belt. When first detected on 1940 July 31, its System II longitude was decreasing at the rate of about 1° per day. Shortly after the middle of August, however, this rate began to diminish and before the end of the month the spot had become almost stationary in System II with, if anything, a tendency to retrograde. This state of affairs lasted until after the middle of September, when the longitude began to diminish again at a rate that reached a maximum just before the middle of October, after which a second deceleration set in, to give a minimum rate of progress near the beginning of November and to be followed by yet another acceleration.

While studying the plotted longitudes of this spot, the author noticed that its oscillations bore at least a superficial resemblance to those of a damped harmonic motion and found upon investigation, that it was possible to make the observations fit such a motion very closely indeed. If the reader will examine the left-hand portion of Fig. 8, which is taken from the longitude chart, he will realise that it was not difficult to detect the wave-like nature of the spot's track, even without the aid of the continuous line, and will note that the amplitude of the waves diminished progressively with the time.

At first the spot, of which the latitude remained sensibly constant throughout its history, was a conspicuous object, though its intensity was subject to some variations. After the middle of September it became generally fainter; but it was successfully followed until December 21, by which time it had become difficult to make out, except in good seeing. Its most interesting feature was, of course, the oscillatory nature of its motion; and as this will be discussed in detail in the theoretical part of this book, where a possible explanation of its behaviour is also suggested, we add no more here but turn to a

OSCILLATING SPOTS

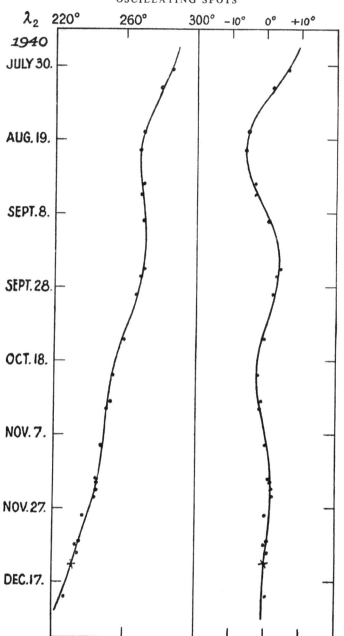

FIGURE 8.—Motion in longitude of the Oscillating Spot of 1940–41.

186

somewhat, though by no means precisely, similar object that was observed during the following apparition.

It was during the latter half of November 1941 that this second object made its appearance. Initially its size, darkness, situation and motion were almost the exact counterparts of the oscillating spot of 1940–41. During the first week in December a deceleration of the rate at which its System II longitude had been decreasing rendered it stationary about December 7, after which it retrograded with diminishing velocity until about December 22, when a further spell of direct motion set in, which lasted until a second minimum longitude was reached towards the end of January 1942. Then a final period of retrogression began, the velocity of which continued to increase until the spot was last certainly seen on March 9. This retrogression was still in operation nearly three weeks later, if two observations obtained at the end of March relate to the same feature —see Fig. 9.

Although the contrast between the final motions of these two spots appears striking enough, the main factor that militates against their being regarded as strictly comparable objects lies in the very definite variations that took place in the latitude of the 1941–42 spot. Starting close to the N. edge of the S.T.B., it moved northwards and about the middle of December was situated near the middle of the S. Tropical Zone, where it remained until, at the earliest, the end of the final week in January. The records of the spot's latitude are unfortunately not very precise, since observers were concentrating mainly upon its excursions in longitude and did not realise at the time the importance of making an equally reliable record of its motion in the other co-ordinate. However, there is some evidence that from the second week in January it moved southwards again until the middle of February, when it began once more to travel towards the north. On February 11 its position was fairly accurately noted as lying just to the south of the middle of the S. Tropical Zone. This northward progress continued during the remainder of the month and there was agreement that on March 2 the spot was fairly close to the S. edge of the S. component of the S.E.B. A week later it seems to have been farther south again; but the records of March 24 and 26, whether or not they refer to the last remnant of this interesting object, are both of a small dark spot or projection at the S. edge of the S.E.Bs.

Now it seems impossible to escape the conclusion that, whatever may have been the prime reason for the oscillation in longitude of the 1941–42 spot, variation of latitude must have contributed to its

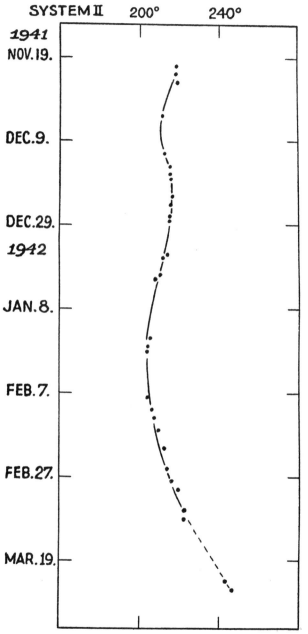

OSCILLATING SPOTS

FIGURE 9.—Motion in longitude of the Oscillating Spot of 1941–42.

188

changes of rotation period; and since the record of the latter is far from being complete or entirely trustworthy, any attempt at the far more difficult task of fitting a significant mathematical curve to the track of this spot on the longitude chart would scarcely seem to justify the labour involved, even though the result should appear superficially to have been successful.

CHAPTER 20

VERY RAPID CHANGES
IN SURFACE FEATURES

In this chapter are described a few instances where changes in form or intensity have been recorded during a single passage of a marking across the disk of Jupiter. That they are rare need occasion no surprise.

Delicate detail is seldom easy to define when it is more than 30° from the central meridian; and so, at a single presentation, a feature may be critically observed for not much longer than the time it takes for the planet to rotate through 60°, which is about $1^h 40^m$. A few cases of change rapid enough to have been recognisable in this interval have been recorded, however, and attention is directed here to three of them, of which two at least bear the stamp of reliability.

During the great revival of the S. Equatorial Belt in the autumn of 1928 very rapid alterations in the forms of markings were suspected on several occasions and on 1928 October 14 Hargreaves, using an 18-inch reflector, recorded quite definitely a change in the aspect of one feature after an interval of not more than about 40 minutes and possibly not exceeding 13 minutes. His two sketches of the same region are here reproduced—Plate X, Fig. 5. The upper picture is part of a 'strip-sketch' of about 150° in length, which he made of the details between the S.E.Bn. and the S.T.B. and which was started at $1^h 30^m$. The region shown was central at about $3^h 12^m$ and the drawing of it was probably started some little time—up to 20 minutes, say—before this and checked at meridian passage. The lower picture was made at $3^h 25^m$. As an experienced observer, well acquainted with Hargreaves' skill and care in planetary draughtsmanship, the author can state that only under the most exceptional circumstances could the curved dark object, projecting upwards between the two white spots in the later drawing, have been missed when it was on the central meridian. The definition was good and there seems to be no room for doubt that a real and striking change was recorded. It is probable that other rapid variations were in progress at the same time, for on that night several portions of the 'strip-sketch' were revised on the first assumption that the originals were erroneous; it now seems more likely that the need for some, at least, of this revision was occasioned by real changes that had actually occurred in the features concerned. It will be remembered that the S.E.B. and its environs were at that time in an abnormally disturbed state.

Going back a quarter of a century, we find (*M.N.*, 63, 704, 1905) the record of a most interesting observation, made by Molesworth in Ceylon. This was also associated with the S. Equatorial Belt but this time the belt as a whole was not in a specially active state. On 1903 December 17 at $7^h 25^m$, local time (G.M.T. $1^h 55^m$), a small white spot, like Satellite I after ingress at transit, was noticed at the S. edge of the S.E.B., not far from the f. shoulder of the Red Spot Hollow. It had not been seen a few minutes earlier, when, if it existed, it must have crossed the central meridian under almost perfect conditions for observation and would surely have been entered in the record of transit times, which Molesworth always kept most assiduously. It quickly formed a deep indentation in the dark belt and by $7^h 30^m$ had become a bright rift, swollen at its p. end and almost joining a white spot at the f. extremity of another rift, which extended into the belt from the Equatorial Zone. Meanwhile the visibility of the object fluctuated considerably; but this was not apparently due to any defect of seeing, as definition is stated to have been excellent throughout the time of observation. The new feature was followed without exhibiting further change, until it was too far from the central meridian to be viewed satisfactorily. There remained no trace of it, however, when the same region was next seen on the evening of December 20.

Our third example is quite well known and dates back to the year 1839, when the following remarkable record was made by Sir James South. It is transcribed here exactly as it is quoted in Rev. T. W. Webb's *Celestial Objects for Common Telescopes*, sixth edition, edited by Espin.

'On June 3, 1839, at $13^h 45^m$ (sidereal time) I saw with my large achromatic, immediately below the lowest [? edge] of the principal belt of Jupiter, a spot larger than I had seen before: it was of a dark colour, but certainly not absolutely black. I estimated it at a fourth of the planet's equatorial diameter. I showed it to some gentlemen who were present: its enormous extent was such that on my wishing to have a portrait of it, one of the gentlemen, who was a good draftsman, kindly undertook to draw me one: whilst I, on the other hand, extremely desirous that its actual magnitude should not rest

191

on estimation, proposed, on account of the scandalous unsteadiness of the large instrument, to measure it tricometrically (*sic*) with my 5-ft. equatorial. Having obtained for my companion the necessary drawing instruments, I went to work, he preparing himself to commence his; on my looking, however, into the telescope of the 5-ft. equatorial, at $13^h 45^m$ (*sic*), I was astonished to find that the large dark spot, except at its eastern and western extremities, had become much whiter than any of the other parts of the planet, and at $14^h 19^m$ these miserable scraps were the only remains of a spot which, but a few minutes before, had extended over at least 22,000 miles.'

One cannot help wondering whether the scientific value of this entry exceeds its literary merit. The author of this book confesses that he has sometimes speculated as to whether the three-letter word, which appears twice in parentheses, might not more appropriately have been written as '*hic*'!

Should there be any who, from an inherent distrust of the reliability of visual planetary observation, would question the accuracy of much of what has been recounted, it is because they have failed to appreciate the nature of the observations upon which the major part of our present knowledge of the behaviour of Jupiter's atmosphere has been founded. It is true that in the past bitter controversy has arisen over the interpretation of visual observations that would ascribe this or that exact form to certain delicate features on the surface of a planet; but how many of the results, that are claimed in this book to have been established, have depended on the detection of minute detail and of these how many have been based upon the form rather than upon the position of such detail? Determination of position provides the data for quantitative treatment; and it is because such treatment has been employed whenever possible, that a high degree of confidence may be placed in so many of the results that have been derived from the visual observation of Jupiter, especially, of course, in the rates of drift of the surprisingly many permanent, temporary and intermittent surface currents, each a characteristic of its own particular latitude. In a few cases, where inferences have been drawn from somewhat meagre or questionable data, the author has been at pains to call attention to the possibility of error; if there are instances of his having omitted to do so, he has certainly failed in his duty to the reader. Let the sceptics, if any remain, study again Chapters 4, 5 and 6, particularly the last, wherein, it is hoped, the nature of the observations and the manner in which they have been reduced have been adequately explained.

As the determination of Rotation Periods has been based entirely upon visual observations, this is an appropriate place to display in a single Table the mean rates of drift and the latitudes of the many surface currents whose existence has been revealed by the observations.

Situation and/or Name of Current	Approx. Latitude	Change of Longitude in 30 days		Rotation Period	No. of Apparitions
		λ_1	λ_2		
	° °	°	°	h m s	
N. Polar Current	+90 to +47		+1	9 55 42	19
N.N.N. Temperate Belt (N.N.N. Temperate Current)	+43		−15	9 55 20	9
N.N. Temperate Belt and Zone (N.N. Temperate Current A)	+40 to +36		+1	9 55 42	35
S. edge of N.N. Temperate Belt (N.N. Temperate Current B)	+35		−78	9 53 55	6
N. edge of N. Temperate Belt and S. part of N. Temp. Zone (N. Temperate Current A)	+33 to +29		+18	9 56 5	24
Middle of N. Temperate Belt (N. Temperate Current B)	+27		−105	9 53 17	6
S. edge of N. Temperate Belt (N. Temperate Current C)	+23	−62		9 49 7	6*
N. Tropical Zone and N. part of N. Equatorial Belt (N. Tropical Current)	+22 to +14		−9	9 55 29	52
Middle of N. Equat. Belt { 1898–1900	?		−6	9 55 32	2 }
1927–1948	+13		−67	9 54 9	7 }

* Not included here are the values for 1880, 1891, and 1892, given to the nearest minute only in a previous table.

Situation and/or Name of Current	Approx. Latitude	Change of Longitude in 30 days		Rotation Period	No. of Apparitions
		λ_1	λ_2		
	° °	°	°	h m s	
S. edge of N. Equatorial Belt and N. part of Equat. Zone (Great Equatorial Current: Northern Branch)	+10 to +3	−4		9 50 24*	47
Middle of Equatorial Zone (Great Equatorial Current: Central Branch)	+3 to −3	−4		9 50 24*	14
N. edge of S. Equatorial Belt and S. part of Equat. Zone (Great Equatorial Current: Southern Branch)	−3 to −10	⎧A −3 ⎨ ⎩B +38		9 50 26* 9 51 21	39⎫ 11⎬
S. edge of S. Equatorial Belt (normal)	−19		−1	9 55 39	13
S. edge of S. Equatorial Belt (Northern Branch of Circulating Current)	−19		+132	9 58 43	6
S. Tropical Zone	−21 to −26		−3	9 55 36	7
The Great Red Spot, 1872–1948	−22		−2	9 55 38	64
N. edge of S. Temperate Belt (Southern Branch of Circulating Current)	−27		−116	9 53 2	5
S. Temperate Belt (S. Temperate Current)	−29		−15	9 55 20	47
S. edge of S. Temperate Belt to S.S. Temperate Zone (S.S. Temperate Current)	−31 to −45		−25	9 55 7	42
S.S.S. Temperate Belt and S. Polar Region	−45 to −90		−8	9 55 30	5

* Although these three mean values are almost identical, in any given year the rates of drift of the different branches of the Great Equatorial Current may be quite independent of one another.

CHAPTER 21

PHOTOGRAPHIC OBSERVATIONS
LIMITATIONS, METHODS AND
ACHIEVEMENTS

To those readers who have been impressed with the enormous advantage possessed by the photographic plate over the human eye in penetrating the depths and elucidating the structure of the starry heavens, but who have not seriously enquired into the reasons for this superiority, it may come as something of a shock to learn that in making records of the surface features of the planets the position is reversed and that the eye is, or has been until very recently, a more sensitive instrument than the camera.

It is not after all surprising that different types of apparatus should be found to give the best results in two widely different lines of research; and both the nature and in general the object of photographic and visual observations are indeed fundamentally different. In the majority of cases the aim of stellar and nebular photography is to obtain a record of very faint objects on a comparatively small scale, whereas in planetary photography the image is comparatively bright and the scale required very large. Two of the chief advantages of the photographic plate as a medium for portraying the stars and nebulae is that its action is cumulative and that thousands of images can be imprinted upon it at the same time. The response of the eye to the stimuli of light quanta is, however, non-cumulative and what cannot be seen in one second cannot usually be seen at all. It is true that the retina becomes gradually more sensitive to faint stimuli during a period of up to a quarter of an hour in the dark; but its response is still practically instantaneous and observers will sometimes speak of having 'glimpsed' a very faint object. If anyone thinks that this is not in accordance with his experience, because the longer he looks at a planet the more detail he can see, he is forgetting the incorrigible habit of the eye to scan the object it is studying and that the accumulation is in his memory and not on his retina.

On the small scale of most stellar photographs the distortion of the images by atmospheric disturbance hardly counts, except inasmuch as the exposure time required for faint stars may be somewhat longer on nights of bad definition. The shutter of the camera can remain open for hours while fainter and fainter objects register themselves, the limit being set only by the ultimate fogging of the plate by the faint but ever present luminosity of the night sky. But in planetary photography we are faced with an entirely different problem. We require the resolution of fine detail on a large scale in an image which, though bright, is not quite bright enough for very rapid exposures with moderate-sized telescopes. If only Mars and Jupiter were just a little brighter or just a little nearer! But should we then be satisfied or should we continue to magnify the images in search of finer and finer detail until once again the old problems were presented?

The factor that militates against the success of any but the shortest exposures is bad definition due to atmospheric turbulence. This is of various kinds but may be broadly divided into two classes, that which confuses fine detail and that which causes the image as a whole to oscillate. In the latter case the eye has little difficulty in holding the detail but the effect on a photographic plate may be as harmful as in the former. In all types of seeing some moments are better than others; the best are generally of amazingly short duration and one of the ways in which the eye of an experienced observer excels is in its ability to seize upon and make the most of these rare opportunities, many of which last for only a second or two. A photographic exposure of the order of five seconds has little chance of being favoured with optimum definition throughout.

There is really a vicious circle at work. We can increase the density of the image on the plate, which would otherwise be underexposed, in three ways:

(1) By lengthening the exposure.
(2) By forming a smaller image of the planet and thereby concentrating the light.
(3) By using a faster photographic emulsion.

We have already seen the disadvantages of the first method. In the second we lose resolution because the granular structure of the silver deposited from the emulsion remains unchanged while the size of the image is reduced; and in the third it is a most unfortunate characteristic of photographic emulsions, not yet successfully overcome by the manufacturers, that increase in speed cannot be obtained without an increase in the size of the silver grains deposited.

There is a fourth method, which lies outside the circle except that, for it to be effective, steadier atmospheric conditions may be required; and that is to employ a larger telescope. Now only a very few of the world's largest instruments are capable of obtaining with an exposure as short as one second a photographic image of Jupiter showing as much detail as can be seen visually with a 12-inch objective. Some beautiful examples of what they can do are reproduced here in the Frontispiece and as Plates XI and XII; but for most of the time their great light-grasp is required for the solution of problems originating far outside the Solar System and it is only rarely that they can be spared to take an occasional planetary photograph. The 24-inch refractor of the Lowell Observatory has done good work in the hands of E. C. Slipher and many of W. H. Wright's beautiful photographs of the planets in light of different wavelengths were obtained with the 36-inch Lick reflector. No doubt the perfection of the atmospheric conditions which is often experienced at both Flagstaff and Mount Hamilton contributed to the success of the exposures of three to four seconds that must have been given; for very exceptional circumstances must surely have prevailed when Slipher obtained the images of which two examples have been reproduced here as Fig. 2 of Plate XIV. Even after the lapse of forty years these still represent what to the visual observer are probably the most realistic photographic likenesses of Jupiter that have ever been achieved. By amateurs with apertures even up to 24 inches little fine detail is likely to be registered, though the more prominent surface features may of course be recorded and even measured on some of the best plates. When delicate detail has to be detected the eye wins every time, except when it has to compete with the world's giants working under the most favourable conditions.

There is one promising technique which in the hands of the expert may give worth-while results with instruments of only moderate aperture. This was developed some years ago by B. Lyot with the object of reducing the effect of the coarseness of the grain that is at present an unavoidable feature of the faster emulsions and was employed with negatives obtained with the 24-inch refractor of the Pic-du-Midi Observatory. The procedure is to utilise a period when the definition is first class to obtain a number of fully exposed and fully developed negatives of the planet; and then to select the best of these and from them to produce a composite picture. This is done in the enlarger by the following method.

Let us suppose that N negatives have been selected and that to make a satisfactory enlargement of one of them would require an exposure time of T seconds. The negatives are placed one after another in the enlarger with the most scrupulous attention to orientation and registration and each is given an exposure of T/N seconds. The result of this process is clearly to build up in the final enlarged image the true planetary details and to average out the effect of the silver grains, whose distribution on the original negatives is fortuitous.

The manipulation required to bring about exact superposition of the images must be very delicate but considerable success has been achieved by Lyot and his associates. In the case of Mars as many as sixteen negatives have been successfully superposed; owing, however, to the time that must elapse between exposures this is an impracticable number for Jupiter, whose more rapid rotation would give rise to too much displacement of the detail if the whole series were to occupy more than about two minutes. For this reason it has been found that the most satisfactory composite pictures of Jupiter are to be obtained from the superposition of four or five separate films. Examples of photographs obtained at the Pic-du-Midi are shown as Figs. 1 and 2 of Plate XIII, the former being the work of Lyot himself and the latter that of H. Camichel.

Another possibility of which we are bound to hear more before long is the application of electronics to planetary photography. In some experiments conducted by the British Broadcasting Corporation, with Hargreaves acting as astronomical adviser, it was found possible to use the image of Jupiter, formed by the 36-inch Yapp reflector of the Royal Greenwich Observatory, to project upon an ordinary television screen an image of the planet having a diameter of from 3 to 4 inches. This image was bright enough to have been photographed with a very short exposure. Jupiter's polar diameter at the time was about 45 seconds of arc and if, say, 120 of the 400-odd scanning lines were involved in forming the picture, the resolution in latitude should have been about 0·4 second, which represents the separating power of a 12-inch objective. In practice, however, accurate spacing of the lines is difficult to attain and the actual resolution would probably have been more of the order of one second. In the other co-ordinate resolution would have depended on the longitudinal extension of the scanning spot and would probably have been similar, the detail registered being thus equivalent to that obtainable from a 5-inch objective. This is hardly good enough; but to increase the number of lines in the image presents no serious technical problem and the control of their spacing and of the dimensions of the scanning spot will doubtless yield quickly to research that is enthusiastic enough to seek refinements beyond those of the commercial needs of the moment, so that we may look forward to an adaptation of this method that may give, perhaps in the near future, photographic resolution equal to that which is optically associated with an aperture of about half that of the objective actually being employed.

The greatest real contribution to our knowledge of Jupiter that has yet been made photographically is undoubtedly due to W. H. Wright, to whose images of the planet, obtained in the light of different wavelengths, reference has already been made. A beautiful series of such photographs, taken in ultra-violet, violet, green, yellow, red and infra-red light, is reproduced here as Plate I. The small spot that is equally dark in all the images is the shadow of a satellite and the dark elliptical ring, which appears in the same latitude as the shadow but on the left-hand side of the disk in the three upper pictures, is the Great Red Spot. It will be noted that this object is extremely dark in the ultra-violet and violet and quite a dark grey in the green; in the yellow its form is less distinct and in the red it is still less easy to identify, while in the infra-red only a trace of it remains. This shows how deficient it was in light of the shorter wavelengths at the time when the photographs were taken and confirms the impression of its ruddy appearance recorded by many visual observers. As opposed to this, the dusky projections that are characteristic of the S. edge of the N. Equatorial Belt have often been described visually as having a bluish tinge. This implied deficiency in the longer wavelengths should cause these features to show up in contrast to the general brightness of the Equatorial Zone in the red and infra-red pictures and this is exactly what we see. In the ultra-violet image the S. edge of the N.E.B. is practically straight; from violet to yellow inclusive there is an indication of a southward hump near the following (right-hand) limb but in the red and infra-red there are three conspicuous projecting humps.

CHAPTER 22

SUGGESTIONS FOR AN OBSERVING PROGRAMME

It is, one fears, too much to hope that one of the really large telescopes will ever be devoted entirely to the study of Jupiter for six months or so each year. The author would like, however, to suggest a compromise, whereby an instrument that could be spared for regular part-time work might embark upon a programme that would offer the maximum return for the hours available.

Let us assume the main object of the programme to be to follow the drifts in longitude of as many spots as possible, any changes in latitude being of course automatically recorded at the same time. If interest were confined to the more prominent features, most of which are comparatively easy to identify and have rotation periods that do not differ widely from those of one or other of the two standard systems, a representative set of photographs taken once a fortnight, or possibly even once every three weeks, should be sufficient for the determination of the mean rotation periods of a fair number of the more long-enduring objects.

Many of the most interesting spots, however, are small, numerous and not readily distinguishable from one another, so that, in the absence of any initial clue as to their rotation periods, it would be almost certainly impossible to identify them after the interval of a fortnight and probably after a week. Suppose, however, that $12\frac{1}{2}$ per cent of the telescope's working time could be spared for Jupiter and that this was allotted so that 25 per cent of the three months on either side of the date of opposition could be utilised. Then, upon certain assumptions that will be explained below, a spell of four consecutive nights' work, alternating with a period of twelve nights when the telescope would be engaged upon some other branch of research, should give results that would be comparable with those that would be obtained by a team of several skilled visual observers working in co-operation with one another.

It would of course be a condition for the success of the programme that the climate should be a favourable one; for the loss of a large fraction of one of the four-nightly observing periods might be disastrous. The assumptions just mentioned are that good images of Jupiter should be obtainable for $2\frac{1}{2}$ hours on either side of the time of meridian passage and that detail on the photographs could be measured with reasonable accuracy at distances up to 30° from the centre of the disk. For during the five hours that would be available for photography each night 180° of longitude would pass the central meridian, which means that the measurable features would extend over 240°. From this it is easy to see that during a successful spell of four nights' observation at least two reliable records of every longitude would be obtained and many regions of the surface would have been photographed three times. When, owing to daylight near quadrature, observing time became curtailed on one side of the meridian, the chances of getting a double record would still be good, even when conditions were favourable for photography for only three hours on each of the four nights.

Let us apply this scheme to an imaginary example and suppose that there is another outbreak of a large number of small, rapidly moving dark spots at the S. edge of the N. Temperate Belt or an upheaval of the S. Equatorial Belt that gives rise to a chain of rapidly retrograding grey spots on the S. component, such as those whose motion was almost certainly missed by visual observers in January 1920 or those which Phillips detected in 1928. However many spots there may be, within reasonable probability, and whatever their motions, pairs of photographs from one of the four-day periods should succeed in establishing identifications and in determining preliminary rates of drift. It is then more than likely that, when the next batch of photographs becomes available, the knowledge already obtained will enable the spots to be identified again, after which more accurate rotation periods will be derived and a general review will have become possible of the changes that have taken place during the interval.

Visual observers may feel some chagrin if the time ever comes when they can no longer claim that in the field of planetary observation and particularly in the study of Jupiter they are still supreme. It may not be yet; but there are signs that the old order may begin to change before long. As an example, the records of visual observers have been most carefully studied in an attempt to locate the source of the recently detected radio 'noise' from Jupiter; but although this study appears to have narrowed down the search to one or two fairly small areas if not to have actually pin-pointed the sources, at least one large telescope has been allotted time for an attempt to throw further light on the subject photographically.

It may require some sensational discovery like this to awaken a wider interest in the possibilities of planetary photography; it is impossible to predict how future generations of astronomers will view its importance or provide for its practical application.

CHAPTER 23

MOTION OF THE
OSCILLATING SPOTS

In the section of this book dealing mainly with the observations a preliminary account of the behaviour of these two objects has already been given. Here we examine in greater detail the unusual motion of the earlier of the two, which appeared as a dark spot, close to the N. edge of the S. Temperate Belt, in the summer and autumn of 1940. As has been stated, its latitude remained sensibly constant; it is the exceedingly close resemblance of its progress in longitude to a damped harmonic motion that is so remarkable and so stimulating to the imagination.

The left-hand side of Fig. 8 shows the actual drift of the spot in System II longitude. The dots represent 24 longitude determinations, all by the author; although a few more were communicated by two other observers, it was decided, for the sake of homogeneity of material, not to include them in the analysis.

In attempting to formulate the motion mathematically it seemed clear from the figure that we should consider some form of damped oscillation, superposed upon a general tendency for the longitude of the spot to decrease. In addition the possible effect of a small constant acceleration was taken into account, so the equation representing the System II longitude λ of the spot at any time t was assumed to be of the form:

$$\lambda = a + bt + ct^2 + de^{-kt} \sin \theta t$$

Since a period not far from 62 days seemed to be indicated, the value $5°.8$ was chosen for θ; it was then possible, after assuming a trial value for k, to determine the four remaining coefficients by the method of least squares. After two or three successive approximations had thus been obtained by varying slightly the values assigned to k and θ, the residuals had become as small as could be expected and the following equation, in which t is the time elapsed in days from 1940 September 8, was adopted as a satisfactory representation of the motion of the spot. It is represented by the continuous line in the left half of Fig. 8.

$$\lambda = 270°\!.8 - 0°\!.274t - 0°\!.00151t^2 + 4°\!.77e^{-0\cdot015t}\sin 5°\!.80t$$

The accompanying table gives the details of the 24 observations, together with their residuals as derived from the above formula and the weights, from 1 to 3, that were assigned to them by the observer according to the conditions under which they were obtained; λ_0 is the observed System II longitude and λ_c that calculated from the formula, while in the column headed $O-C$ are the residuals in the sense observed minus calculated. Two other observations were made in conditions that were such as to warrant complete rejection, although one of them would have given a zero residual. The column $O-C'$ is explained below.

Date		t	λ_0	λ_c	$O-C$	W	$O-C'$
1940							
July	31	−39	286°	285°.4	+0°.6	3	+1°.3
Aug.	5	34	280	280·7	−0·7	1	+0·9
	17	22	271	270·9	+0·1	2	+0·5
	22	17	269	268·9	+0·1	3	−0·3
	31	8	270	269·0	+1·0	3	−0·1
Sept.	3	−5	269	269·6	−0·6	3	−1·7
	10	+2	271	271·3	−0·3	3	−0·9
	23	15	271	270·1	+0·9	1	+0·4
	25	17	269	269·4	−0·4	2	−0·9
	30	22	266	266·8	−0·8	3	−0·5
Oct.	12	34	260	258·9	+1·1	3	+0·2
	22	44	254	253·6	+0·4	2	−0·1
	29	51	252	250·9	+1·1	1	+1·3
	31	53	250	250·3	−0·3	3	+0·2
Nov.	10	63	247	247·7	−0·7	3	+0·8
	19	72	244	244·6	−0·6	1	+0·8
	20	73	244	244·2	−0·2	3	+1·2
	22	75	244	243·3	+0·7	2	+1·8
	24	77	243	242·3	+0·7	3	+1·7
	29	82	237	239·4	−2·4	2	−2·3
Dec.	6	89	235	235·0	0·0	3	−0·7
	7	90	233	234·3	−1·3	3	−2·2
	9	92	234	233·0	+1·0	3	−0·1
	21	+104	227	225·1	+1·9	3	−0·7

The probable error of timing the spot across Jupiter's central meridian is estimated to have been about two minutes, of which the equivalent in longitude is $1°.2$. Of the 24 residuals all but three are less than this and only one exceeds it by a factor 2. It will be conceded, therefore, that the formula is an adequate representation of the drift of the spot in longitude.

Now the last term in the expression, taken by itself, represents accurately a damped harmonic oscillation, such as is performed by a pendulum or the prong of a tuning-fork, vibrating in a resisting medium. This motion is depicted graphically in the right-hand half of Fig. 8, where the continuous line represents the oscillating term alone and the dots have been plotted from the numbers in the λ_0 column of the table, after their values have been diminished by the algebraic sum of the first three terms in the formula. The steady decrease in the amplitude of the vibration, while the period remains constant, is well shown.

Another example that closely resembles a damped harmonic oscillation is the vertical motion of the egg in the saline solution, referred to in the last chapter, while it is recovering its equilibrium after a displacement from its natural level; and it is easy to show that, if a large floating body were to oscillate harmonically up and down, like the egg, in the atmosphere of a rotating planet, its angular drift in longitude, consistent with the conservation of angular momentum, would also be harmonic. The difference in phase between the two vibrations would depend upon the value of the damping coefficient and would be $90°$ in the absence of damping.

Just such an effect has been observed, namely the damped harmonic oscillation in longitude of a dark spot at the upper surface of Jupiter's cloud layer during the closing months of 1940. Is it too much to suggest that the cause may have operated in a manner not essentially different from the behaviour of the egg?

If the reader now decides that the author is riding his hobby-horse unbridled into the realms of pure fantasy, he need expect no apology. Such an excursion can be very exhilarating and may be beneficial rather than harmful, provided that the critical faculty remains alert and as eager to expose the many fallacies with which it may be presented as to grasp and extend a promising idea. We will imagine, then, that for some reason, which we must admit seems basically improbable and for which the author cannot offer even a tentative explanation, the density of portions of the outer crust of the solid body of the planet is very slightly lower than that of the fluid immediately above it. Let us further suppose that not many days before 1940 July 31 a large fragment of this lighter material became detached from its surroundings and began to float upwards, the combined thermal and mechanical effects being responsible for the appearance of a dark spot in the cloud layer vertically above it. Had this indeed been the case, the observed motion of the spot in longitude, as the solid fragment rose and fell in the atmosphere, would have conformed very closely to a curve similar to one of those shown in Fig. 8.

But a point arises that had escaped the author at the time when his original paper on the Oscillating Spot was published.* It can be seen, if either of the curves is examined carefully, or it can be calculated by differentiating the formula, that the maximum rate at which the longitude of the spot decreased does not appear to have been attained until about August 5, which was therefore the date when the oscillating solid reached its lowest point; so at the time of its discovery it must have been going downwards. But according to the present very tentative theory this will not do at all. The time when it broke away from the bottom of the atmospheric ocean must have coincided with the time when the rate of decrease of longitude was at a maximum and this should have occurred at least a day or two before the spot was first recorded. Moreover, the initial maximum rate of rotation of the spot about Jupiter's axis should have corresponded almost exactly with the rotation period of the solid body of the planet.

Now in deriving our representative formula the damping coefficient k and the period were originally obtained by guesswork and the constants employed in the expression are by no means the only ones that will give an adequate representation of the observations. Quite recently the author, in the hope of being able to remove the above inconsistencies from a theory which, whatever its shortcomings, is not likely to be accused of being inartistic, has been examining alternative solutions and has found that the following expression, in which the damping coefficient has been slightly modified and the period increased to 72 days, satisfies most of his requirements:

$$\lambda = 271°6 - 0°310t - 0°00118t^2 + 5°30e^{-0.018t} \sin 5°0t$$

The residuals, which are those given in the last column of the previous table, are not quite so satisfactory from the minimum square point of view; but only 8 out of the 24 are greater than the

* M.N., 101, 70, 1941.

probable error and not one exceeds it by a factor as great as 2. As this is in any case almost too good to be true, it cannot be argued that either formula is the more representative; but the second expression, in which $d\lambda/dt$ has its greatest negative value on July 29 or 30, about a day and a half before the spot was first observed, interprets the data in a manner that is no longer inconsistent with the author's speculations.

No separate diagram to represent the new formula is given here, as the curves would resemble so closely those already exhibited in Fig. 8. The main difference to be noted would be that the point of inflexion, shown above as occurring on August 5 at $t = -34$, has been displaced to the top of the curve at approximately $t = -40\cdot5$; towards the bottom, where the difference in phase would become rather obvious, the amplitude has become so small that the observations would be represented almost as well by a straight line as by either of the curves. The daily rate of change of longitude of the spot at $t = -40\cdot5$, given by the second expression, is now readily found to be $-1^\circ\!.176$ in System II or $871^\circ\!.446$ relative to the vernal equinox of Jupiter. It has been pointed out that if the above ideas are tenable this figure, which corresponds to a rotation period of $9^h\ 54^m\ 52^s\!.4$, should represent the angular velocity of the solid surface of the planet; and on referring back to Chapter 27 we find that the shorter of the two periods which Reese derived from the longitudes of the origins of the S.E.B. eruptions was $9^h\ 54^m\ 52^s\!.54$. What a truly remarkable agreement! As a matter of fact it is not especially wonderful, because in seeking suitable coefficients for the terms in the second formula, one of the author's chief concerns was to see how close an agreement he could obtain with Reese; but that he was able to achieve such a striking result without having to do violence either to the observations or to his theory does at least show that the ideas involved are not mutually inconsistent.

Again using the revised formula and remembering that the latitude of the spot was about $-26°$, we can further ascertain that the greatest height to which the solid would have risen in the atmosphere would have been some 65 kilometres above its starting level and that this would have been attained at about $t = -4\cdot5$; we also find that 18 days before this it would have achieved its maximum upward velocity of almost exactly 3 cm. per second. The force per unit mass, diminishing but nevertheless continuous, that is necessary for generating so small a velocity in 18 days is, even initially, almost incredibly minute; and it can be calculated that it would result from a difference in density between the solid and the surrounding fluid,

207

when the former began its upward career, which was of the order of only 3×10^{-7} per cent. So perhaps our original assumption, that the solid fragment was less dense than the fluid initially in contact with it, was not so fantastic after all; for if the two densities were normally about equal, it would take very little in the way of a temperature change in either of the media to bring about the necessary instability.

The author started this chapter in a highly sceptical frame of mind; but things seem to have worked out more plausibly than he had anticipated, with the result that, at the stage that has now been reached, he has very nearly succeeded in converting himself to a belief in his own theory!

Towards the end of the original paper an attempt was made to employ some of the figures already evaluated to calculate the compressibility of the fluid medium, on the assumption that the compressibility of the solid was negligible in comparison. The resulting value was so many orders of magnitude too small that it seemed at first to invalidate the whole theory. It was not long, however, before Professor Bridgman, in a private communication to the author, pointed out that the compressibility of the solid was by no means negligible but might be of the same order of magnitude as that of the fluid at the high pressures involved. The improbably low value had therefore been obtained upon an erroneous assumption and was of no significance as a criterion for the validity of the rest of the paper.

After the rather astonishing way in which the motion of the first of the two spots had proved amenable to analysis, the longitude of the oscillating spot of 1941–42, having aroused our enthusiasm by making a most promising start, failed lamentably to fulfil our expectations. As was pointed out in the chapter on the observations, this second spot underwent considerable changes in latitude and it seems clear that the type of motion in longitude that it would have exhibited if its latitude had remained constant must have been appreciably modified by its response to the variation that was observed in the latter co-ordinate. Had its progress towards the equator been sensibly uniform instead of apparently capricious, the observations might have been worth analysing; for it would have been natural to expect a systematic acceleration in longitude and the bottom part of the curve of Fig. 9 does strongly suggest that it is asymptotic to a parabola. As it is, even if an interesting result were obtained from an analysis of the motion, there would always be the feeling that it had probably been vitiated by the effect of these latitude changes; so up to date the author has not seriously attempted the rather heavy arithmetic that would be involved, though he would be the last to discourage any of his readers from trying to wrest some interesting information from the curve. The record of the observations, which have been described in the appropriate chapter, ought to be reliable enough to rule out the possibility that the diminution of latitude was uniform; but the agreement about the position of the spot in the S. Tropical Zone between Hargreaves and the author, who were the two principal observers, was not always as good as it should have been, since both were concentrating chiefly upon the accurate determination of its longitude. So perhaps a loophole remains for those who would like to assume a strong parabolic component in the motion.

PART III

THE SATELLITES

CHAPTER 24

A GENERAL SURVEY OF THE
SATELLITE SYSTEM

More than a dozen satellites of Jupiter are known. Of these the four brightest were the first new members of the Solar System to be discovered after the invention of the telescope. They are commonly known as the Galilean satellites, for until recently it was generally accepted that Galileo had the undisputed right to be named their discoverer. Historical research, however, has made out a strong case for the claim that Simon Marius should be regarded as an independent discoverer, in that he had probably been observing these objects at the same time as Galileo or even a month or two earlier and certainly, he stated, before the news had reached him that they had already been observed. The evidence is far from conclusive and those who are interested are advised to study the full and impartial investigation that has been undertaken by J. H. Johnson and published in the *B.A.A. Journal*, Vol. 41, p. 164, together with the comments by P. Pagnini on p. 415 of the same volume.

The satellites are generally designated by Roman numerals, which have been allotted to the four brightest in order of their distances from Jupiter and to the remainder in order of discovery. The following names have also been given to the Galilean satellites:

I	Io	III	Ganymede
II	Europa	IV	Callisto

All the rest of the satellites owe their discovery to photography with the exception of V. On 1892 September 9 Professor E. E. Barnard, using the 36-inch refractor of the Lick Observatory and hiding the disk of Jupiter behind a 'bar' in the field of view to minimise the glare of the planet, detected a minute point of light closely following it. This quickly disappeared; but on the following night he saw it again and was very soon able to confirm that it was indeed a new satellite. Its period of revolution of about $11^h 57\frac{1}{2}^m$ is almost exactly two hours longer than the rotation period of the planet.

The following Table gives the principal data concerning the satellites.

No.	Name	Mean Distance from Jupiter: 10^3 km. Jupiter	radii.	Period days	Diameter km.	Magnitude	Year of Discovery
14	Adrastea	134	1.76	0.30	40	20	1979
5	Amalthea	181	2.55	0.49	200	13	1892
15	—	222	3.11	0.67	70	19	1980
1	Io	422	5.95	1.77	3638	4.8	1610
2	Europa	671	9.47	3.55	3126	5.2	1610
3	Ganymede	1070	15.10	7.15	5276	4.5	1610
4	Callisto	1880	26.60	16.70	4848	5.5	1610
13	Leda	11,100	156	240	15	20	1974
6	Himalia	11,470	161	251	100	14	1904
10	Lysithea	11,710	164	260	20	19	1938
7	Elara	11,740	165	260	30	16	1904
12	Ananke	20,700	291	617	20	19	1951
11	Carme	22,350	314	692	20	18	1938
8	Pasiphaë	23,300	327	735	20	19	1908
9	Sinope	23,700	333	758	20	18	1914

The diameters of the smaller satellites are naturally uncertain. The orbits of the outer satellites (XIII to IX) are so strongly affected by the Sun that they are not even approximately circular, and, as noted, the last four have retrograde motion. Amateurs will be concerned only with the Galileans, which will be referred to here by their numbers rather than their names.

I and III would be easy naked-eye objects but for the proximity of the brilliant planet; II and IV would be near the limit but might be seen on occasions, as their brightness is variable. Claims that one or more of them have been actually detected without optical aid have been numerous but are of more interest to the physiologist than to the astronomer. A very modest pair of binoculars will reveal them all, while to the beginner who has acquired a 2-inch or 3-inch telescope, their varying configurations as they revolve about the planet provide a constant source of pleasure and interest.

Since the orbits of these satellites lie very nearly in the plane of Jupiter's equator, which is inclined at only slightly more than 3° to its orbit around the Sun, they alternately transit across the disk of the planet and are occulted behind it during the course of each revolution, IV being the only one that is far enough away from the primary ever to pass clear of the disk, which it does when the declination of the Earth as seen from Jupiter numerically exceeds about 2°7. Their shadows cast by the Sun upon the face of the planet near their times of transit are fascinating to watch, since they appear as little black dots that may be seen with a very small instrument. But perhaps the most intriguing of all the more common satellite phenomena are their eclipses by the shadow of Jupiter. During the months before opposition a satellite that is apparently approaching occultation may be seen, while it is still some distance from the planet, to fade gradually to extinction as it passes into the shadow. Neglecting the small effect due to the penumbra, the process of fading in the case of I, which is the most rapidly moving of the four, lasts just about three minutes; it takes somewhat longer for the others, according to distance from the primary and to the obliquity at which they enter the cone of shadow. After opposition the shadow is on the other side of Jupiter and a satellite that has been occulted will first reappear at some distance from the terminator as a minute point, rapidly growing in brilliance as it emerges into full sunlight. The distances of III and IV are large enough, however, to allow both disappearance and reappearance at a single eclipse to be observed on the same side of the planet, either before or after occultation, when the angle between the Earth and Sun as seen from Jupiter is sufficiently large; and very occasionally, when the planet is within a day or two of quadrature, this may just manage to take place in the case of II. As in the case of the transits and occultations, IV is the only one far enough away from Jupiter to experience fairly long periods without eclipse; the others must pass through the shadow at every revolution.

Predictions of all these phenomena are to be found in the National Ephemerides.

There is an interesting relation between the motions of I, II and III. It is easily found from the Table that the mean daily motion of I added to twice that of III is equal to three times the motion of II. This relation is exact and permanent and can be shown mathematically to be a necessary consequence of the mutual gravitational perturbations between the three satellites. More precisely, if θ_1, θ_2, θ_3 are respectively the mean longitudes of the satellites in their orbits, measured from some initial radius through the centre of Jupiter, then

$$\theta_1 + 2\theta_3 - 3\theta_2 = 180°$$

This ensures that all three satellites can never exhibit similar phenomena at the same time; for instance, II and III may be seen in transit together across the disk but, if so, I is occulted by the

215

planet. It may seem curious that this close relation is quite unaffected by the presence of IV, whose motion is incommensurable with that of the other three.

Further reference to the Table will show that I and II are both a little smaller than the Moon, though I is more massive, while III and IV are appreciably larger, the former being of almost exactly the same size as the planet Mercury but of less than half its mass. All of them are subject to variations in brightness with a range of the order of half a stellar magnitude, which must be attributed to the changing aspect of light and dusky markings as different parts of their surfaces are presented to the observer. It is now generally believed that each of these satellites turns always the same face towards Jupiter in a manner precisely similar to that in which the Moon perpetually shows the same side to the Earth. This was to be expected from the theory of tidal friction; but Phillips was among the first to obtain satisfactory observational confirmation that at any rate in the case of III it is very approximately true. In telescopes of moderate aperture one of the effects of incompletely resolved surface detail is to deform the apparent image of a satellite so that it no longer looks quite round. Between 1915 and 1921, Phillips, using a very fine 8-inch refractor, systematically measured the position angle of the major axis of III whenever it appeared to be elongated. After the figures he obtained had been plotted against the longitude of the satellite in its orbit, measured from superior geocentric conjunction, a convincing correlation was apparent, indicating that at any given point in its orbit the aspect it presented was the same. Recently, however, Professor W. H. Steavenson, from a study with the aid of considerable optical power of the distribution of light and shade over the little disk of III during transit, has found that, while the pattern always repeats itself over short periods, it is different now from that which used to be presented forty years ago. If this is indeed the case, we have an interesting example of a tidal force having nearly but not quite completed its work of synchronising the periods of rotation and revolution of a satellite.

The differences in albedo or average surface brightness per unit area of these four little moons is well brought out when they are projected during transit against the luminous background of Jupiter's disk; so too is the fact, obvious in some photographs but not easy to detect visually, probably on account of contrast with the darkness of the sky, that the brightness of the planet falls off very considerably towards the limb. Near the beginning of transit even quite a small telescope will show them all as bright spots against a somewhat

216

darker background; but they are soon lost to view and neither I nor II will be seen again, except with an instrument of fair size, until they reappear as little bright spots as the end of transit approaches. With sufficient aperture, however, I, after becoming temporarily lost, will almost always reappear as a small dusky spot when its background is one of the zones; but if it is projected against one of the belts it may be quite difficult to recapture it. As a dusky spot it will frequently look far from round, indicating that it is a configuration of dusky markings on the satellite and not the little disk as a whole that we see projected against the brighter surface behind it. The only one of the four that usually remains bright during the whole of its passage across the planet is II; even this may not be easy to detect against the middle of a bright zone, but against any of the belts it should always be easy to see. Except when seen against the more dusky parts of the Polar Regions, III and IV always turn into conspicuous dark spots not long after the beginning of transit; IV indeed may do so astonishingly soon after it has entered upon the disk. Either may be mistaken for a shadow when none is present on the disk for comparison, yet they both appear bright when really close to the limb. When the shadow of either III or IV is in transit at the same time as the satellite, the latter will normally be seen to be not really black, though there is at least one record of the two looking equally dark. At such times the shadow will appear larger than the satellite, providing a demonstration that the penumbra contributes a good deal to the apparent size of the shadow—compare the sizes of III and its shadow exhibited in the Frontispiece. The actual umbra of the shadow of IV is in fact quite a small dot in the middle of a very much larger area of penumbra, yet the whole of what we see looks quite black.

Further reference to the actual surface markings of the satellites is deferred, for a reason that will emerge later, until an account has been given of another series of beautiful phenomena that may be presented by these little bodies.

CHAPTER 25

MUTUAL OCCULTATIONS AND ECLIPSES, SURFACE FEATURES AND FREAK OBSERVATIONS

Once in every 5·93 years the plane of Jupiter's equator passes through the Sun and for some months around these times both the Sun and the Earth remain close to the planes of the orbits of the Galilean satellites. Their apparent motions, if viewed from either the Earth or the Sun, are then back and forth very nearly in straight lines and it is clear that there may be frequent instances of one of them passing in front of another. When this happens as seen from the Earth, we call the phenomenon a mutual occultation. When, however, this effect would be visible to a hypothetical observer on the Sun, it is evident that the shadow of the nearer satellite must fall upon the farther and that an observer on the Earth, to whom the satellites may not even appear particularly close together, will see the latter totally, annularly, partially or perhaps only penumbrally eclipsed.

There is nothing especially striking about the mutual occultations. A point of interest, however, is that they bring out well the difference in the albedos of the satellites, especially when the darker is the nearer to the observer. On 1932 February 18 W. H. Steavenson, Phillips and the author independently made drawings of a partial occultation of I by IV. The author's instrument was a 12¼-inch reflector, while Phillips was presumably employing his 8-inch refractor. Their two sketches were almost identical, an outstanding feature of both being a very dark border to IV along that part of the limb only that overlapped I. Doubtless this effect was partly a physiological one, induced by contrast, and partly due to failure of the apertures used to give adequate optical resolution; for the drawing of Steavenson, whose instrument was a 20-inch reflector, showed no trace of the dark border.

Predictions of mutual occultations are a convenience but by no means a necessity; for the observer, if already at work, can see the

218

satellites approaching one another and be ready to note any interesting occurrence. This is by no means the case, however, with the eclipses; for the shadows cannot be seen approaching and the apparent distance between the eclipsed and eclipsing satellites may be quite large; moreover, the duration of the phenomenon may be so short that it could easily be missed by an observer who was actually at the telescope but was not especially looking for it. Neither of these phenomena is at all rare when the Sun and Earth are near the plane of Jupiter's equator; yet it is stated in the sixth edition of Rev. T. W. Webb's well-known *Celestial Objects for Common Telescopes*, brought up to date and published in 1917 by Rev. T. E. Espin, that there was only one case on record of an observation of the eclipse of one satellite by the shadow of another. The first to be found in the annals of the British Astronomical Association is dated 1920 February 8, when C. S. Saxton, having observed with a 3-inch refractor an occultation of I by III, calculated that the corresponding eclipse, if it occurred at all, would be seen about 20 minutes later. Although thick mist came up and at times obscured the satellites, a partial clearance at the expected time, $14^h 22^m$, revealed II, III and IV but gave no sign of I. Detailed calculation subsequently showed that an eclipse had indeed taken place which, though not total, reached its maximum phase at $14^h 23^m7$ and that at $14^h 22^m$ I had already lost a considerable proportion of its normal light.

On the approach of the apparition of 1926, when the next series of mutual phenomena was due, the author was fortunately able to arouse the interest of Major A. E. Levin, who was then the Director of the B.A.A. Computing Section and who entered enthusiastically into the task of supplying predictions of the eclipses for intending observers. The author had the privilege of being the first to verify one of these predictions observationally; the occasion was the eclipse of I by II which took place at sunrise on 1926 May 22 and which was well seen in spite of the brightness of the sky; so Levin was greeted by a telegram on his breakfast table congratulating him on the accuracy of his calculations. Thereafter many mutual eclipses were observed, one being recorded by no less than seven independent observers; prediction and observation continued through the apparition of 1931–32 and though predictions were published for the next series they were fewer in number and poorly observed, for the simple reason that the orbit planes passed through the Sun while Jupiter was near conjunction between the none too favourable apparitions of 1937 and 1938. Then came World War II; and it is not surprising that no definite predictions were available for 1943–44. Some of the times of mutual heliocentric conjunction of the satellites were published, however, to enable observers to be on the watch at possible times and on one occasion the author was thus led to observe the fading of II in the shadow of I.

The war has been over now for more than a decade and the reader may be wondering why no further predictions have been forthcoming, even for the apparition of 1955–56. For many years the work involved in performing the calculations that are necessary for the production of the National Ephemerides, such as the *Nautical Almanac*, *Connaissance des Temps* and the *American Ephemeris*, has been apportioned by international agreement between the various offices concerned and the results have been pooled for publication, with enormous saving of time and labour. Among the first steps in the work involved in predicting any phenomena connected with Jupiter's satellites, which is the responsibility of the *Connaissance des Temps*, is the derivation of certain fundamental quantities, such as the varying eccentricities and nodes of the orbits, direct from Professor R. A. Sampson's monumental 'Tables of the Four Great Satellites', a labour not to be lightly undertaken by the amateur computer. When Levin began his calculations for 1926 the results of this preliminary work were already available and were supplied to him through the courtesy of the office of the *Connaissance des Temps*, whence similar data were also kindly furnished for the next two series of mutual phenomena. It was natural, therefore, that the necessary information failed to materialise in 1944; but what has happened since? Inquiries notwithstanding, no completely satisfactory explanation has been given to the author, who can therefore only suspect either non-cooperation, apathy or perhaps, more charitably, shortage of staff combined with pressure of other work in the quarters concerned. Quite apart from the importance of the means they provide for checking accurately the positions of the satellites, it seems a tragedy that these beautiful phenomena should be allowed to take place 'unheralded', for they would certainly not pass 'unsung'. All enthusiastic observers of Jupiter and many more besides will agree that it is imperative that predictions should be published for the passage of the Sun through the planes of the orbits in 1961.

There are two possible satellite configurations that the author would deem it a great privilege to behold, though he probably never will, as the second at any rate must be rather rare. One is the eclipse of one satellite by the shadow of another when the former and both the shadows are in transit across the disk of Jupiter. What will happen, of course, is that the two shadows will coalesce and the satellite will turn black—surely a delightful spectacle! The second configuration is illustrated in Fig. 10. The outer satellite, which must be either III or IV and which has not yet begun to be eclipsed by

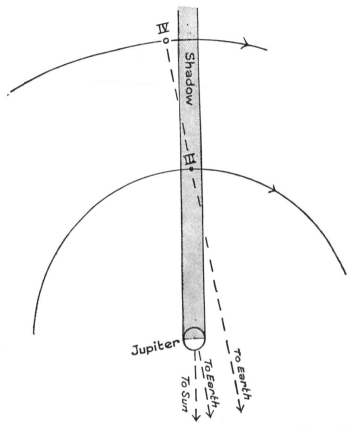

FIGURE 10. —Diagram illustrating the occultation of a visible satellite by one that is eclipsed in Jupiter's shadow.

Jupiter's shadow, is occulted by one of the others which is already immersed in the shadow and is therefore completely invisible. For such an occultation to be total the inner satellite must be III and the outer IV. What a very satisfying observation it would be!

So far no reference has been made to the actual appearance of a satellite whilst it is suffering eclipse in the shadow of another, excepting the brief remark that the eclipses may be either total, annular, partial or penumbral. The only total eclipses that are certainly possible are those of I and II in the shadow of III. But when I and II are both near the remotest parts of their orbits, the umbra of I, according to the figures in the Table given above, may reach 97 per cent of the diameter of II at the mean distances of the satellites; so under exceptionally favourable circumstances a brief totality might be achieved. In general, however, any eclipse that is nearly central will be annular, the umbra having a smaller diameter than the disk of the eclipsed satellite and therefore being surrounded by a luminous ring that is affected by penumbra only. If the two satellites are in widely distant parts of their orbits the length of the cone of true shadow may be less than the distance between them, when even a central eclipse will be entirely penumbral as is the case when an annular eclipse of the Sun is seen from the Earth. Partial eclipses need no comment, as every central eclipse is preceded and followed by a partial phase.

No doubt the reader has been imagining the pretty picture that must be presented by one of these tiny disks with a still smaller black dot in the middle of it; but no such appearance has yet been recorded. The following is an account of what Phillips saw with his 8-inch refractor, power 600, on 1931 December 19, when an annular eclipse of I by II took place. The prediction was that the penumbra of II would be affecting I from $4^h 5^m$ to $4^h 10^m$ and the umbra from $4^h 6^m$ to $4^h 9^m$, the diameter of the umbra being 0·36 of the diameter of the disk of I. At $4^h 5^m5$ the shadow was seen to be taking a bite out of the f. side of I that was unexpectedly large, probably because it was mostly the penumbra that was responsible for the appearance. At about $4^h 7^m$, however, the shadow of II seemed to have disappeared as such, the disk of I being dimmed all over with only a *suspected* darkening at the centre. Then at about $4^h 8^m$ the shadow (penumbra?) reappeared as a large dark notch on the p. side of I and by $4^h 8^m5$ I was normal again, indicating that the predicted times were about 1 minute late. On 1926 July 13 during an annular eclipse of II by I Saxton, using an $8\frac{1}{2}$-inch reflector, had had an almost identical experience; but to him, at the time when the annulus should have been seen, the whole disk of the satellite seemed to expand somewhat and to assume a woolly appearance. On 1931 December 26, the occasion being an annular eclipse of III by the shadow of I, the author, using a power of 300 on his $12\frac{1}{4}$-inch reflector, had the same experience as Phillips without even suspecting annularity; the approach and departure of the shadow were fairly obvious but no trace of a central dark spot was detected.

These three entirely independent observations, together with the absence of any report that an annulus had been seen, ultimately led the author to carry out some experiments that might throw a little light on the optics of the phenomenon. He mounted on a dark background some small white paper disks, varying in diameter from 1·35 to 4·6 millimetres, putting little black spots on them in different positions to represent shadows in transit. These were placed at a distance of 118 metres, where they could be viewed with a 3-inch refractor that had been provided with a number of stops for reducing the aperture by stages to 1 inch. The angular subtense of the disks ranged from 2″·4 to 8″·0 and it was found that no trace of the central dark spot could be seen on a 4″·2 disk with less than 2 inches aperture nor on an 8″·0 disk with 1 inch. As the critical aperture was exceeded the change in appearance was rapid; for instance, the central spot on the 8″·0 disk was always obvious at first glance with a 1·2-inch stop. Since resolving power is directly proportional to aperture, it is readily seen from these results that an 8-inch telescope is inadequate to reveal the annularity of an eclipse of a satellite whose diameter is 1 second of arc. Should the disk be larger than this, the critical aperture is of course reduced in proportion. The failure of Phillips and Saxton to see what was optically impossible is immediately explained; it appears, however, that the author's 12¼-inch should have been equal to the task of resolving the annulus, particularly as the satellite was III and probably subtended nearly 1″·5, but in this case the diameter of the umbra was only 0·15 of that of the satellite and as definition was not good enough to warrant a magnifying power of more than 300 the conditions must have fallen far short of laboratory standards.

On two other paper disks, each subtending 5″·9, fairly conspicuous pencil marks were made, on one a cross and on the other a crescent. The grey cross was never seen properly as such, even with the full 3 inches aperture; sometimes it appeared simply as a *black* dot like the others and sometimes as a black dot with faint grey wisps indicating the arms of the cross. With less than 3 inches nothing other than a dot was ever seen and with 1·4 inches the disk was blank. The crescent, seen with the full aperture, was sometimes a small, plain black dot, at others a straight streak, condensed in the middle, or just a shapeless elongated patch. Its curvature was never apparent. Thus, while it must be admitted that an unsymmetrically placed marking can cause a satellite to appear elongated instead of round even when there is no resolution, the above experimentally obtained figures indicate that a dusky marking on a disk subtending 1 second

223

of arc cannot be seen at all as such with less than 8½ inches of aperture and cannot be properly resolved even with 18 inches.

Perhaps it will now have become clear to the reader why the discussion concerning the markings on the satellites was deferred until their mutual eclipses had been described. A large number of drawings purporting to show the surface details of the satellites have been made, mostly with apertures ranging from 6 to 10 inches; yet it is doubtful whether many of the draughtsmen, reading this, will admit even to themselves that they have represented something that wasn't there. By all means let us have drawings of the satellites made with apertures of 25 inches and over on the rare occasions when the definition justifies it; but the possessor of anything much smaller is strongly urged to carry out similar experiments for himself before he publishes the impressions he receives at the eyepiece of an instrument that is almost certainly inadequate.

It would be a good thing in any case for the experiments to be repeated independently. The author's apparatus was somewhat crude and it seems probable that more realistic effects would be produced if the little circles with the black dots were photographed and displayed as transparencies with diffused light behind them. Under ideal laboratory conditions it might perhaps transpire that the critical apertures given here would bear a reduction of, say, 20 per cent; but certainly not much more. A word of warning, however. Do not mount 1-inch disks a mile away and try to observe them in daylight. The author has tried it; and the vilest seeing he has ever experienced in his observatory could hardly have competed with what he viewed through a mile of the lowest stratum of our atmosphere, even when the sky was overcast.

These conclusions as to the inadequacy of telescopes of moderate size to show correctly the details of the surface markings on the satellites do not necessarily apply when a satellite is in transit across the disk of Jupiter. Further experiments, with similar pencil markings on little white disks and on a large sheet of white paper, indicate that not more than half the resolving power is required to reveal the details in the latter case than in the former. If then the satellite becomes projected against a background that causes its circular outline to disappear, considerably more confidence may be placed in any impression received of the form of dusky surface detail than if the satellite were being viewed normally against the dark sky.

In conclusion attention may be called to the quite considerable number of 'freak' observations of the satellites, many of them quite fantastic, to which reference may be found in the literature of the subject. There is, for example, the instance, supposed to be attested by three independent observers, of a satellite which having entered in transit upon the disk of Jupiter was seen 12 or 13 minutes later outside the limb, where it remained for 4 minutes before it suddenly vanished. It is difficult to understand the attitude of mind that in the absence of a wholly satisfactory explanation would hold that this may really have happened; if, however, the reader should ever hear such an appearance being ascribed to the intervention of a Jovian Joshua, he is strongly advised not to argue! If through the telescope we seem to behold the normal laws of physics being disregarded, the appearance is nearly always to be explained as some optical anomaly, most probably due to one of the vagaries of 'bad seeing'. A spurious effect that has been recorded many times is the apparent doubling of a satellite shadow. This comes within the experience of the author, who on 1942 February 19 saw for minutes together, except for a few seconds now and again of normal vision, the shadow of I, which was traversing the S.E.Bn., accompanied by a companion shadow, quite as sharply defined and only a little smaller, on the S.E.Bs. The observation was a perfectly good one of course—of the image formed in the focal plane of the telescope, not of what was happening on Jupiter—and can doubtless be explained on some such hypothesis as that a pocket or column of air of different temperature from its surroundings was tending to become stabilised over a portion of the 12¼-inch mirror. There can be little doubt but that an electric fan, directed to the proper quarter, would have dispelled the illusion. Yet in *M.N.*, 34, 65, 1874, may be found an attempt to determine a lower limit to the depth of Jupiter's atmosphere that was based upon certain elongated appearances of the shadows that had been previously recorded. That the effect of a shadow was supposed to have been observable and indeed observed after penetrating to a depth of several thousand miles—surely an obvious impossibility, whatever the nature of the atmosphere—does not seem to have deterred the council of no less august a body than the Royal Astronomical Society from accepting this flight of fancy for publication.

The moral of the last paragraph is twofold: that until he has learnt to discredit almost automatically the maxim 'seeing is believing' the man at the little end of a telescope will never become a trustworthy planetary observer; and further that, before basing his calculations upon observation, the theorist should have a reasonably good idea of the nature of the actual observational procedure and of the pitfalls lying in wait for the unwary observer.

REDUCTION OF LATITUDE MEASUREMENTS

(See p. 49)

In Fig. 11a the ellipse represents a section of Jupiter by the plane through its poles and the Earth, the eccentricity being greatly exaggerated for the sake of clarity.

C is the centre of Jupiter, CA the equatorial radius, B one of the poles and CE the direction of the Earth. On the actual planet $CA=1 \cdot 0714CB$.

Then the angle $ACE=D_\oplus$.

P is a point on the central meridian which has been measured for latitude and CD is the semi-diameter conjugate to CE.

PQ is drawn parallel to DC to meet CE at Q.

Then, since the tangent to the ellipse at D is parallel to CE, the measured quantity that has been called $\sin \theta$ is PQ/DC.

Let the tangent to the ellipse at P meet CA produced at T and CB produced at t and let N, M be the feet of the perpendiculars from P, Q respectively on CA.

In this figure the angle PCA is the Zenocentric latitude β and the angle PtC the Zenographical latitude β''.

In Fig. 11b the ellipse of Fig. 11a has been projected to its auxiliary circle, dashed letters representing the projections of the undashed counterparts of those points which have been displaced in the process.

Now all lengths perpendicular to CA in Fig. 11a are increased by a factor $1 \cdot 0714$ in Fig. 11b, therefore

$$1 \cdot 0714 = \frac{QM'}{QM}$$

$$= \frac{\tan ACE'}{\tan ACE}$$

$$= \frac{\text{angle } ACE'}{\text{angle } ACE}$$

since angle ACE cannot exceed $3^\circ 07$; whence angle $ACE'=D'_\oplus$.

227

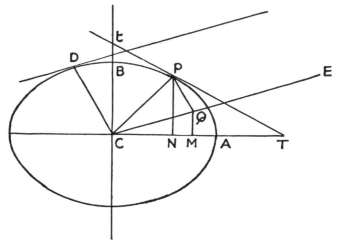

FIGURE 11a

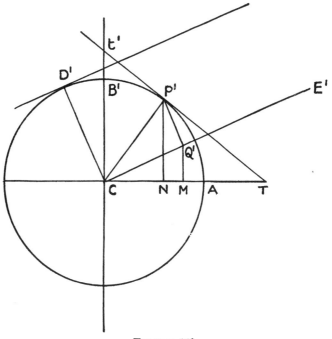

FIGURE 11b
228

Again, since the ratios of parallel lengths are unchanged by projection,

$$\sin \theta = \frac{PQ}{DC}$$
$$= \frac{P'Q'}{D'C}$$
$$= \frac{P'Q'}{P'C}$$

so that angle $P'CQ' = \theta$ and angle $P'CN = \theta + D'_\oplus = \beta'$.

But
$$\tan \beta = \frac{PN}{CN}$$
$$= \frac{1}{1 \cdot 0714} \frac{P'N}{CN}$$
$$= 0 \cdot 9333 \frac{P'N}{CN}$$
$$= 0 \cdot 9333 \tan \beta'$$

and
$$\tan \beta'' = \frac{CT}{Ct}$$
$$= 1 \cdot 0714 \frac{CT}{Ct'}$$
$$= 1 \cdot 0714 \tan Ct'T$$
$$= 1 \cdot 0714 \tan P'CN$$
$$= 1 \cdot 0714 \tan \beta'.$$

It seems a pity to the author that the reduction has ever been carried beyond the determination of β', the eccentric angle of the point P. This angle has surely more significance from the terrestrial point of view than either the zenocentric or the zenographical latitude; for $\sin \beta'$ is the measured fraction of the polar semi-diameter when the Earth is overhead at Jupiter's equator and it is $\sin \beta'$ that has been used for the ordinates of Plate II, the pictorial representation of latitude changes. Perhaps the expression 'Eccentric Latitude' or, more simply, 'Mean Jovian Latitude' would not have been out of place had it been employed to indicate this angle. As it is, one fears it is too late for a change to be desirable.

APPENDIX II

THE TIME OF APPARENT CENTRAL MERIDIAN PASSAGE OF A SATELLITE SHADOW IN TRANSIT

(See p. 54)

The progress of the shadow is assumed to be parallel to Jupiter's equator and in Fig. 12 the circle represents the parallel of latitude along which it is travelling.

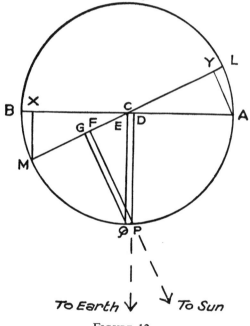

FIGURE 12

C is the centre of the circle, AB and LM being the diameters which are perpendicular to the directions of the Earth and Sun respectively as seen from Jupiter.

X is the foot of the perpendicular from M on AB, and Y that of

230

the perpendicular from A on LM. D is the middle point of AX and E the middle point of YM.

Perpendiculars from D and C to AB meet the circle at P and Q respectively, while the feet of the perpendiculars from P and Q to LM are F and G.

The semicircle $LAPQM$ is illuminated by the Sun; but a shadow travelling round it from L to M will be visible only from A to M, A being on the limb and M on the terminator as seen from the Earth.

Letting T_1 and T_2 be the times when the visible shadow begins transit at A and ends at M, we assume that the progress of its projection along LM is uniform, so that at time $(T_1+T_2)/2$ the projection of the shadow is at E. But as seen from the Earth the shadow clearly appears to be central when it reaches P, at which time its projection is at F.

It is evident, therefore, that the time of arrival of the shadow at P is $(T_1+T_2)/2$ plus a correction representing the time taken by the projection to travel from E to F. This correction is the fraction EF/YM of T_2-T_1.

Since the angle ACL $(=DPF=\theta)$ cannot exceed $12°$, the arc PQ is very approximately a straight line parallel to AB; therefore

$$\frac{r(1-\cos\theta)}{2}=CE=DC=PQ=FG \sec\theta,$$

where r is the radius of the circle.

Also $CG=r\sin\theta$ and $YM=r(1+\cos\theta)$,

whence $\quad \dfrac{EF}{YM}=\dfrac{CG-CE-FG}{YM}$

$$=\frac{\sin\theta-\frac{1}{2}(1-\cos\theta)-\frac{1}{2}(1-\cos\theta)\cos\theta}{1+\cos\theta}$$

$$=\frac{\sin\theta}{1+\cos\theta}-\frac{1}{2}(1-\cos\theta)$$

$$=\tan\theta/2-\frac{1}{2}+\frac{1}{2}\cos\theta.$$

The correction to $(T_1+T_2)/2$ is therefore

$$(T_2-T_1)(0\cdot5\cos\theta+\tan\theta/2-0\cdot5).$$

JUPITER

MOVEMENT OF THE CENTRAL MERIDIAN
SYSTEM I

m	0ʰ	1ʰ	2ʰ	3ʰ	4ʰ	5ʰ	6ʰ	7ʰ	8ʰ	9ʰ	10ʰ	11ʰ
0	0.0	36.6	73.2	109.7	146.3	182.9	219.5	256.1	292.7	329.2	5.8	42.4
1	0.6	37.2	73.8	110.4	146.9	183.5	220.1	256.7	293.3	329.8	6.4	43.0
2	1.2	37.8	74.4	111.0	147.5	184.1	220.7	257.3	293.9	330.5	7.0	43.6
3	1.8	38.4	75.0	111.6	148.2	184.7	221.3	257.9	294.5	331.1	7.6	44.2
4	2.4	39.0	75.6	112.2	148.8	185.3	221.9	258.5	295.1	331.7	8.3	44.8
5	3.0	39.6	76.2	112.8	149.4	186.0	222.5	259.1	295.7	332.3	8.9	45.4
6	3.7	40.2	76.8	113.4	150.0	186.6	223.1	259.7	296.3	332.9	9.5	46.1
7	4.3	40.8	77.4	114.0	150.6	187.2	223.8	260.3	296.9	333.5	10.1	46.7
8	4.9	41.5	78.0	114.6	151.2	187.8	224.4	260.9	297.5	334.1	10.7	47.3
9	5.5	42.1	78.6	115.2	151.8	188.4	225.0	261.6	298.1	334.7	11.3	47.9
10	6.1	42.7	79.3	115.8	152.4	189.0	225.6	262.2	298.7	335.3	11.9	48.5
11	6.7	43.3	79.9	116.5	153.0	189.6	226.2	262.8	299.4	335.9	12.5	49.1
12	7.3	43.9	80.5	117.1	153.6	190.2	226.8	263.4	300.0	336.5	13.1	49.7
13	7.9	44.5	81.1	117.7	154.3	190.8	227.4	264.0	300.6	337.2	13.7	50.3
14	8.5	45.1	81.7	118.3	154.9	191.4	228.0	264.6	301.2	337.8	14.3	50.9
15	9.1	45.7	82.3	118.9	155.5	192.1	228.6	265.2	301.8	338.4	15.0	51.5
16	9.8	46.3	82.9	119.5	156.1	192.7	229.2	265.8	302.4	339.0	15.6	52.1
17	10.4	46.9	83.5	120.1	156.7	193.3	229.9	266.4	303.0	339.6	16.2	52.8
18	11.0	47.6	84.1	120.7	157.3	193.9	230.5	267.0	303.6	340.2	16.8	53.4
19	11.6	48.2	84.7	121.3	157.9	194.5	231.1	267.7	304.2	340.8	17.4	54.0
20	12.2	48.8	85.4	121.9	158.5	195.1	231.7	268.3	304.8	341.4	18.0	54.6
21	12.8	49.4	86.0	122.5	159.1	195.7	232.3	268.9	305.5	342.0	18.6	55.2
22	13.4	50.0	86.6	123.2	159.7	196.3	232.9	269.5	306.1	342.6	19.2	55.8
23	14.0	50.6	87.2	123.8	160.3	196.9	233.5	270.1	306.7	343.3	19.8	56.4
24	14.6	51.2	87.8	124.4	161.0	197.5	234.1	270.7	307.3	343.9	20.4	57.0
25	15.2	51.8	88.4	125.0	161.6	198.1	234.7	271.3	307.9	344.5	21.1	57.6
26	15.9	52.4	89.0	125.6	162.2	198.8	235.3	271.9	308.5	345.1	21.7	58.2
27	16.5	53.0	89.6	126.2	162.8	199.4	235.9	272.5	309.1	345.7	22.3	58.9
28	17.1	53.7	90.2	126.8	163.4	200.0	236.6	273.1	309.7	346.3	22.9	59.5
29	17.7	54.3	90.8	127.4	164.0	200.6	237.2	273.7	310.3	346.9	23.5	60.1
30	18.3	54.9	91.5	128.0	164.6	201.2	237.8	274.4	310.9	347.5	24.1	60.7
31	18.9	55.5	92.1	128.6	165.2	201.8	238.4	275.0	311.6	348.1	24.7	61.3
32	19.5	56.1	92.7	129.3	165.8	202.4	239.0	275.6	312.2	348.7	25.3	61.9
33	20.1	56.7	93.3	129.9	166.4	203.0	239.6	276.2	312.8	349.4	25.9	62.5
34	20.7	57.3	93.9	130.5	167.1	203.6	240.2	276.8	313.4	350.0	26.5	63.1
35	21.3	57.9	94.5	131.1	167.7	204.2	240.8	277.4	314.0	350.6	27.2	63.7
36	21.9	58.5	95.1	131.7	168.3	204.9	241.4	278.0	314.6	351.2	27.8	64.3
37	22.6	59.1	95.7	132.3	168.9	205.5	242.0	278.6	315.2	351.8	28.4	65.0
38	23.2	59.7	96.3	132.9	169.5	206.1	242.7	279.2	315.8	352.4	29.0	65.6
39	23.8	60.4	96.9	133.5	170.1	206.7	243.3	279.8	316.4	353.0	29.6	66.2
40	24.4	61.0	97.6	134.1	170.7	207.3	243.9	280.5	317.0	353.6	30.2	66.8
41	25.0	61.6	98.2	134.7	171.3	207.9	244.5	281.1	317.6	354.2	30.8	67.4
42	25.6	62.2	98.8	135.4	171.9	208.5	245.1	281.7	318.3	354.8	31.4	68.0
43	26.2	62.8	99.4	136.0	172.5	209.1	245.7	282.3	318.9	355.4	32.0	68.6
44	26.8	63.4	100.0	136.6	173.2	209.7	246.3	282.9	319.5	356.1	32.6	69.2
45	27.4	64.0	100.6	137.2	173.8	210.3	246.9	283.5	320.1	356.7	33.2	69.8
46	28.0	64.6	101.2	137.8	174.4	211.0	247.5	284.1	320.7	357.3	33.9	70.4
47	28.7	65.2	101.8	138.4	175.0	211.6	248.1	284.7	321.3	357.9	34.5	71.0
48	29.3	65.8	102.4	139.0	175.6	212.2	248.8	285.3	321.9	358.5	35.1	71.7
49	29.9	66.5	103.0	139.6	176.2	212.8	249.4	285.9	322.5	359.1	35.7	72.3
50	30.5	67.1	103.6	140.2	176.8	213.4	250.0	286.6	323.1	359.7	36.3	72.9
51	31.1	67.7	104.3	140.8	177.4	214.0	250.6	287.2	323.7	0.3	36.9	73.5
52	31.7	68.3	104.9	141.4	178.0	214.6	251.2	287.8	324.4	0.9	37.5	74.1
53	32.3	68.9	105.5	142.1	178.6	215.2	251.8	288.4	325.0	1.5	38.1	74.7
54	32.9	69.5	106.1	142.7	179.2	215.8	252.4	289.0	325.6	2.2	38.7	75.3
55	33.5	70.1	106.7	143.3	179.9	216.4	253.0	289.6	326.2	2.8	39.3	75.9
56	34.1	70.7	107.3	143.9	180.5	217.0	253.6	290.2	326.8	3.4	40.0	76.5
57	34.8	71.3	107.9	144.5	181.1	217.7	254.2	290.8	327.4	4.0	40.6	77.1
58	35.4	71.9	108.5	145.1	181.7	218.3	254.8	291.4	328.0	4.6	41.2	77.8
59	36.0	72.6	109.1	145.7	182.3	218.9	255.5	292.0	328.6	5.2	41.8	78.4
60	36.6	73.2	109.7	146.3	182.9	219.5	256.1	292.7	329.2	5.8	42.4	79.0

JUPITER

MOVEMENT OF THE CENTRAL MERIDIAN
SYSTEM II

m	0ʰ	1ʰ	2ʰ	3ʰ	4ʰ	5ʰ	6ʰ	7ʰ	8ʰ	9ʰ	10ʰ	11ʰ
0	0.0	36.3	72.5	108.8	145.1	181.3	217.6	253.8	290.1	326.4	2.6	38.9
1	0.6	36.9	73.1	109.4	145.7	181.9	218.2	254.4	290.7	327.0	3.2	39.5
2	1.2	37.5	73.7	110.0	146.3	182.5	218.8	255.0	291.3	327.6	3.8	40.1
3	1.8	38.1	74.3	110.6	146.9	183.1	219.4	255.7	291.9	328.2	4.4	40.7
4	2.4	38.7	74.9	111.2	147.5	183.7	220.0	256.3	292.5	328.8	5.0	41.3
5	3.0	39.3	75.5	111.8	148.1	184.3	220.6	256.9	293.1	329.4	5.7	41.9
6	3.6	39.9	76.2	112.4	148.7	184.9	221.2	257.5	293.7	330.0	6.3	42.5
7	4.2	40.5	76.8	113.0	149.3	185.5	221.8	258.1	294.3	330.6	6.9	43.1
8	4.8	41.1	77.4	113.6	149.9	186.1	222.4	258.7	294.9	331.2	7.5	43.7
9	5.4	41.7	78.0	114.2	150.5	186.8	223.0	259.3	295.5	331.8	8.1	44.3
10	6.0	42.3	78.6	114.8	151.1	187.4	223.6	259.9	296.1	332.4	8.7	44.9
11	6.6	42.9	79.2	115.4	151.7	188.0	224.2	260.5	296.7	333.0	9.3	45.5
12	7.3	43.5	79.8	116.0	152.3	188.6	224.8	261.1	297.4	333.6	9.9	46.1
13	7.9	44.1	80.4	116.6	152.9	189.2	225.1	261.7	298.0	334.2	10.5	46.7
14	8.5	44.7	81.0	117.2	153.5	189.8	226.0	262.3	298.6	334.8	11.1	47.4
15	9.1	45.3	81.6	117.9	154.1	190.4	226.6	262.9	299.2	335.4	11.7	48.0
16	9.7	45.9	82.2	118.5	154.7	191.0	227.2	263.5	299.8	336.0	12.3	48.6
17	10.3	46.5	82.8	119.1	155.3	191.6	227.8	264.1	300.4	336.6	12.9	49.2
18	10.9	47.1	83.4	119.7	155.9	192.2	228.5	264.7	301.0	337.2	13.5	49.8
19	11.5	47.7	84.0	120.3	156.5	192.8	229.1	265.3	301.6	337.8	14.1	50.4
20	12.1	48.4	84.6	120.9	157.1	193.4	229.7	265.9	302.2	338.5	14.7	51.0
21	12.7	49.0	85.2	121.5	157.7	194.0	230.3	266.5	302.8	339.1	15.3	51.6
22	13.3	49.6	85.8	122.1	158.3	194.6	230.9	267.1	303.4	339.7	15.9	52.2
23	13.9	50.2	86.4	122.7	159.0	195.2	231.5	267.7	304.0	340.3	16.5	52.8
24	14.5	50.8	87.0	123.3	159.6	195.8	232.1	268.3	304.6	340.9	17.1	53.4
25	15.1	51.4	87.6	123.9	160.2	196.4	232.7	268.9	305.2	341.5	17.7	54.0
26	15.7	52.0	88.2	124.5	160.8	197.0	233.3	269.6	305.8	342.1	18.3	54.6
27	16.3	52.6	88.8	125.1	161.4	197.6	233.9	270.2	306.4	342.7	18.9	55.2
28	16.9	53.2	89.4	125.7	162.0	198.2	234.5	270.8	307.0	343.3	19.6	55.8
29	17.5	53.8	90.1	126.3	162.6	198.8	235.1	271.4	307.6	343.9	20.2	56.4
30	18.1	54.4	90.7	126.9	163.2	199.4	235.7	272.0	308.2	344.5	20.8	57.0
31	18.7	55.0	91.3	127.5	163.8	200.0	236.3	272.6	308.8	345.1	21.4	57.6
32	19.3	55.6	91.9	128.1	164.4	200.7	236.9	273.2	309.4	345.7	22.0	58.2
33	19.9	56.2	92.5	128.7	165.0	201.3	237.5	273.8	310.0	346.3	22.6	58.8
34	20.5	56.8	93.1	129.3	165.6	201.9	238.1	274.4	310.6	346.9	23.2	59.4
35	21.2	57.4	93.7	129.9	166.2	202.5	238.7	275.0	311.3	347.5	23.8	60.0
36	21.8	58.0	94.3	130.5	166.8	203.1	239.3	275.6	311.9	348.1	24.4	60.6
37	22.4	58.6	94.9	131.1	167.4	203.7	239.9	276.2	312.5	348.7	25.0	61.3
38	23.0	59.2	95.5	131.8	168.0	204.3	240.5	276.8	313.1	349.3	25.6	61.9
39	23.6	59.8	96.1	132.4	168.6	204.9	241.1	277.4	313.7	349.9	26.2	62.5
40	24.2	60.4	96.7	133.0	169.2	205.5	241.8	278.0	314.3	350.5	26.8	63.1
41	24.8	61.0	97.3	133.6	169.8	206.1	242.4	278.6	314.9	351.1	27.4	63.7
42	25.4	61.6	97.9	134.2	170.4	206.7	243.0	279.2	315.5	351.7	28.0	64.3
43	26.0	62.3	98.5	134.8	171.0	207.3	243.6	279.8	316.1	352.4	28.6	64.9
44	26.6	62.9	99.1	135.4	171.6	207.9	244.2	280.4	316.7	353.0	29.2	65.5
45	27.2	63.5	99.7	136.0	172.2	208.5	244.8	281.0	317.3	353.6	29.8	66.1
46	27.8	64.1	100.3	136.6	172.9	209.1	245.4	281.6	317.9	354.2	30.4	66.7
47	28.4	64.7	100.9	137.2	173.5	209.7	246.0	282.2	318.5	354.8	31.0	67.3
48	29.0	65.3	101.5	137.8	174.1	210.3	246.6	282.8	319.1	355.4	31.6	67.9
49	29.6	65.9	102.1	138.4	174.7	210.9	247.2	283.5	319.7	356.0	32.2	68.5
50	30.2	66.5	102.7	139.0	175.3	211.5	247.8	284.1	320.3	356.6	32.8	69.1
51	30.8	67.1	103.3	139.6	175.9	212.1	248.4	284.7	320.9	357.2	33.5	69.7
52	31.4	67.7	104.0	140.2	176.5	212.7	249.0	285.3	321.5	357.8	34.1	70.3
53	32.0	68.3	104.6	140.8	177.1	213.3	249.6	285.9	322.1	358.4	34.7	70.9
54	32.6	68.9	105.2	141.4	177.7	213.9	250.2	286.5	322.7	359.0	35.3	71.5
55	33.2	69.5	105.8	142.0	178.3	214.6	250.8	287.1	323.3	359.6	35.9	72.1
56	33.8	70.1	106.4	142.6	178.9	215.2	251.4	287.7	323.9	0.2	36.5	72.7
57	34.4	70.7	107.0	143.2	179.5	215.8	252.0	288.3	324.5	0.8	37.1	73.3
58	35.1	71.3	107.6	143.8	180.1	216.4	252.6	288.9	325.2	1.4	37.7	73.9
59	35.7	71.9	108.2	144.4	180.7	217.0	253.2	289.5	325.8	2.0	38.3	74.6
60	36.3	72.5	108.8	145.1	181.3	217.6	253.8	290.1	326.4	2.6	38.9	75.2

Appendix IV

CRITICAL TABLES
for the conversion of
Change of Longitude in Thirty Days to Rotation Period

A. System I

Change of Longitude in 30 days	Rotation Period 9ʰ 48ᵐ	Change of Longitude in 30 days	Rotation Period 9ʰ 49ᵐ	Change of Longitude in 30 days	Rotation Period 9ʰ 50ᵐ	Change of Longitude in 30 days	Rotation Period 9ʰ 51ᵐ	Change of Longitude in 30 days	Rotation Period 9ʰ 52ᵐ
−112°4	0ˢ	−67°5	0ˢ	−22°7	0ˢ	+21°9	0ˢ	+66°3	0ˢ
111·7	1	66·8	1	22·0	1	22·6	1	67·0	1
110·9	2	66·0	2	21·3	2	23·3	2	67·8	2
110·2	3	65·3	3	20·5	3	24·1	3	68·5	3
109·4	4	64·5	4	19·8	4	24·8	4	69·3	4
108·7	5	63·8	5	19·0	5	25·6	5	70·0	5
107·9	6	63·0	6	18·3	6	26·3	6	70·7	6
107·2	7	62·3	7	17·5	7	27·1	7	71·5	7
106·4	8	61·5	8	16·8	8	27·8	8	72·2	8
105·7	9	60·8	9	16·0	9	28·5	9	73·0	9
104·9	10	60·0	10	15·3	10	29·3	10	73·7	10
104·2	11	59·3	11	14·6	11	30·0	11	74·4	11
103·4	12	58·5	12	13·8	12	30·8	12	75·2	12
102·7	13	57·8	13	13·1	13	31·5	13	75·9	13
101·9	14	57·0	14	12·3	14	32·2	14	76·7	14
101·2	15	56·3	15	11·6	15	33·0	15	77·4	15
100·4	16	55·5	16	10·8	16	33·7	16	78·1	16
99·7	17	54·8	17	10·1	17	34·5	17	78·9	17
98·9	18	54·1	18	9·3	18	35·2	18	79·6	18
98·2	19	53·3	19	8·6	19	36·0	19	80·4	19
97·4	20	52·6	20	7·9	20	36·7	20	81·1	20
96·7	21	51·8	21	7·1	21	37·4	21	81·8	21
95·9	22	51·1	22	6·4	22	38·2	22	82·6	22
95·2	23	50·3	23	5·6	23	38·9	23	83·3	23
94·4	24	49·6	24	4·9	24	39·7	24	84·0	24
93·7	25	48·8	25	4·1	25	40·4	25	84·8	25
92·9	26	48·1	26	3·4	26	41·1	26	85·5	26
92·2	27	47·3	27	2·6	27	41·9	27	86·3	27
91·4	28	46·6	28	1·9	28	42·6	28	87·0	28
90·7	29	45·8	29	1·2	29	43·4	29	87·7	29
89·9	30	45·1	30	− 0·4	30	44·1	30	88·5	30
89·2	31ˢ	44·4	31	+ 0·3	31	44·8	31	+89·2	
−88·4		−43·6		+ 1·1		+45·6			

In critical cases ascend.

A. System I (continued)

Change of Longitude in 30 days	Rotation Period 9ʰ 48ᵐ	Change of Longitude in 30 days	Rotation Period 9ʰ 49ᵐ	Change of Longitude in 30 days	Rotation Period 9ʰ 50ᵐ	Change of Longitude in 30 days	Rotation Period 9ʰ 51ᵐ
−88°4		−43°6		+ 1°1		+45°6	
	32		32		32		32
87·7		42·9		1·8		46·3	
	33		33		33		33
86·9		42·1		2·5		47·1	
	34		34		34		34
86·2		41·4		3·3		47·8	
	35		35		35		35
85·4		40·6		4·0		48·5	
	36		36		36		36
84·7		39·9		4·8		49·3	
	37		37		37		37
83·9		39·1		5·5		50·0	
	38		38		38		38
83·2		38·4		6·3		50·8	
	39		39		39		39
82·5		37·6		7·0		51·5	
	40		40		40		40
81·7		36·9		7·8		52·2	
	41		41		41		41
81·0		36·2		8·5		53·0	
	42		42		42		42
80·2		35·4		9·2		53·7	
	43		43		43		43
79·5		34·7		10·0		54·5	
	44		44		44		44
78·7		33·9		10·7		55·2	
	45		45		45		45
78·0		33·2		11·5		56·0	
	46		46		46		46
77·2		32·4		12·2		56·7	
	47		47		47		47
76·5		31·7		12·9		57·4	
	48		48		48		48
75·7		30·9		13·7		58·2	
	49		49		49		49
75·0		30·2		14·4		58·9	
	50		50		50		50
74·2		29·4		15·2		59·7	
	51		51		51		51
73·5		28·7		15·9		60·4	
	52		52		52		52
72·7		28·0		16·7		61·1	
	53		53		53		53
72·0		27·2		17·4		61·9	
	54		54		54		54
71·2		26·5		18·1		62·6	
	55		55		55		55
70·5		25·7		18·9		63·4	
	56		56		56		56
69·7		25·0		19·6		64·1	
	57		57		57		57
69·0		24·2		20·4		64·8	
	58		58		58		58
68·2		23·5		21·1		65·6	
	59		59		59		59
67·5		22·7		21·9		66·3	
	60		60		60		60
−66·8		−22·0		+22·6		+67·0	

In critical cases ascend.

B. System II

Change of Longitude in 30 days	Rotation Period 9ʰ 52ᵐ	Change of Longitude in 30 days	Rotation Period 9ʰ 53ᵐ	Change of Longitude in 30 days	Rotation Period 9ʰ 54ᵐ	Change of Longitude in 30 days	Rotation Period 9ʰ 55ᵐ
−162°6		−118°3		−74°1		−30°1	
161·9	0ˢ	117·6	0ˢ	73·4	0ˢ	29·4	0ˢ
161·1	1	116·8	1	72·7	1	28·7	1
160·4	2	116·1	2	71·9	2	27·9	2
159·6	3	115·3	3	71·2	3	27·2	3
158·9	4	114·6	4	70·5	4	26·5	4
158·2	5	113·9	5	69·7	5	25·7	5
157·4	6	113·1	6	69·0	6	25·0	6
156·7	7	112·4	7	68·3	7	24·3	7
155·9	8	111·7	8	67·5	8	23·5	8
155·2	9	110·9	9	66·8	9	22·8	9
154·5	10	110·2	10	66·1	10	22·1	10
153·7	11	109·4	11	65·3	11	21·3	11
153·0	12	108·7	12	64·6	12	20·6	12
152·2	13	108·0	13	63·9	13	19·9	13
151·5	14	107·2	14	63·1	14	19·2	14
150·8	15	106·5	15	62·4	15	18·4	15
150·0	16	105·8	16	61·7	16	17·7	16
149·3	17	105·0	17	60·9	17	17·0	17
148·5	18	104·3	18	60·2	18	16·2	18
147·8	19	103·6	19	59·5	19	15·5	19
147·1	20	102·8	20	58·7	20	14·8	20
146·3	21	102·1	21	58·0	21	14·0	21
145·6	22	101·3	22	57·2	22	13·3	22
144·9	23	100·6	23	56·5	23	12·6	23
144·1	24	99·9	24	55·8	24	11·8	24
143·4	25	99·1	25	55·0	25	11·1	25
142·6	26	98·4	26	54·3	26	10·4	26
141·9	27	97·7	27	53·6	27	9·6	27
141·2	28	96·9	28	52·8	28	8·9	28
140·4	29	96·2	29	52·1	29	8·2	29
139·7	30	95·5	30	51·4	30	7·5	30
138·9	31	94·7	31	50·6	31	6·7	31
138·2	32	94·0	32	49·9	32	6·0	32
137·5	33	93·3	33	49·2	33	5·3	33
136·7	34	92·5	34	48·4	34	4·5	34
136·0	35	91·8	35	47·7	35	3·8	35
135·3	36	91·0	36	47·0	36	3·1	36
134·5	37	90·3	37	46·2	37	2·3	37
133·8	38	89·6	38	45·5	38	1·6	38
133·0	39	88·9	39	44·8	39	0·9	39
132·3	40	88·1	40	44·0	40	−0·1	40
131·6	41	87·4	41	43·3	41	+0·6	41
130·8	42	86·6	42	42·6	42	1·3	42
−130·1	43	−85·9	43	−41·9	43	+2·0	43

In critical cases ascend.

236

Change of Longitude in 30 days	Rotation Period 9ʰ 52ᵐ	Change of Longitude in 30 days	Rotation Period 9ʰ 53ᵐ	Change of Longitude in 30 days	Rotation Period 9ʰ 54ᵐ	Change of Longitude in 30 days	Rotation Period 9ʰ 55ᵐ
−130°.1		−85°.9		−41°.9		+2°.0	
	44		44		44		44
129.4		85.2		41.1		2.8	
	45		45		45		45
128.6		84.4		40.4		3.5	
	46		46		46		46
127.9		83.7		39.7		4.2	
	47		47		47		47
127.1		83.0		38.9		5.0	
	48		48		48		48
126.4		82.2		38.2		5.7	
	49		49		49		49
125.7		81.5		37.5		6.4	
	50		50		50		50
124.9		80.7		36.7		7.2	
	51		51		51		51
124.2		80.0		36.0		7.9	
	52		52		52		52
123.4		79.3		35.3		8.6	
	53		53		53		53
122.7		78.5		34.5		9.3	
	54		54		54		54
122.0		77.8		33.8		10.1	
	55		55		55		55
121.2		77.1		33.1		10.8	
	56		56		56		56
120.5		76.3		32.3		11.5	
	57		57		57		57
119.8		75.6		31.6		12.3	
	58		58		58		58
119.0		74.9		30.9		13.0	
	59		59		59		59
118.3		74.1		30.1		13.7	
	60		60		60		60
−117.6		−73.4		−29.4		+14.5	

Change of Longitude in 30 days	Rotation Period 9ʰ 56ᵐ	Change of Longitude in 30 days	Rotation Period 9ʰ 57ᵐ	Change of Longitude in 30 days	Rotation Period 9ʰ 58ᵐ	Change of Longitude in 30 days	Rotation Period 9ʰ 59ᵐ
+13°.7		+57°.4		+101°.0		+144°.4	
	0ˢ		0ˢ		0ˢ		0ˢ
14.5		58.2		101.7		145.1	
	1		1		1		1
15.2		58.9		102.4		145.9	
	2		2		2		2
15.9		59.6		103.2		146.6	
	3		3		3		3
16.6		60.3		103.9		147.3	
	4		4		4		4
17.4		61.1		104.6		148.0	
	5		5		5		5
18.1		61.8		105.3		148.8	
	6		6		6		6
18.8		62.5		106.1		149.5	
	7		7		7		7
19.6		63.3		106.8		150.2	
	8		8		8		8
20.3		64.0		107.5		150.9	
	9		9		9		9
21.0		64.7		108.2		151.6	
	10		10		10		10
21.7		65.4		109.0		152.4	
	11		11		11		11
22.5		66.2		109.7		153.1	
	12		12		12		12
23.2		66.9		110.4		153.8	
	13		13		13		13
23.9		67.6		111.1		154.5	
	14		14		14		14
24.7		68.3		111.9		155.2	
	15		15		15		15
25.4		69.1		112.6		156.0	
	16		16		16		16
26.1		69.8		113.3		156.7	
	17		17		17		17
26.9		70.5		114.0		157.4	
	18		18		18		18
27.6		71.2		114.8		158.1	
	19		19		19		19
28.3		72.0		115.5		158.9	
	20		20		20		20
29.0		72.7		116.2		159.6	
	21		21		21		21
29.8		73.4		116.9		160.3	
	22		22		22		22
30.5		74.2		117.7		161.0	
	23		23		23		23
+31.2		+74.9		+118.4		+161.7	

In critical cases ascend.

Change of Longitude in 30 days	Rotation Period	Change of Longitude in 30 days	Rotation Period	Change of Longitude in 30 days	Rotation Period	Change of Longitude in 30 days	Rotation Period
9^h 52^m		9^h 57^m		9^h 58^m		9^h 59^m	
+31°2		+74°9		+118°4		+161°7	
32·0	24	75·6	24	119·1	24	162·5	24
32·7	25	76·3	25	119·8	25	163·2	25
33·4	26	77·1	26	120·6	26	163·9	26
34·1	27	77·8	27	121·3	27	164·6	27
34·9	28	78·5	28	122·0	28	165·4	28
35·6	29	79·2	29	122·7	29	166·1	29
36·3	30	80·0	30	123·4	30	166·8	30
37·1	31	80·7	31	124·2	31	167·5	31
37·8	32	81·4	32	124·9	32	168·2	32
38·5	33	82·1	33	125·6	33	169·0	33
39·2	34	82·9	34	126·3	34	169·7	34
40·0	35	83·6	35	127·1	35	170·4	35
40·7	36	84·3	36	127·8	36	171·1	36
41·4	37	85·0	37	128·5	37	171·8	37
42·2	38	85·8	38	129·2	38	172·6	38
42·9	39	86·5	39	130·0	39	173·3	39
43·6	40	87·2	40	130·7	40	174·0	40
44·3	41	87·9	41	131·4	41	174·7	41
45·1	42	88·7	42	132·1	42	175·4	42
45·8	43	89·4	43	132·9	43	176·2	43
46·5	44	90·1	44	133·6	44	176·9	44
47·2	45	90·8	45	134·3	45	177·6	45
48·0	46	91·6	46	135·0	46	178·3	46
48·7	47	92·3	47	135·7	47	179·0	47
49·4	48	93·0	48	136·5	48	179·8	48
50·2	49	93·7	49	137·2	49	180·5	49
50·9	50	94·5	50	137·9	50	181·2	50
51·6	51	95·2	51	138·6	51	181·9	51
52·3	52	95·9	52	139·4	52	182·6	52
53·1	53	96·6	53	140·1	53	183·4	53
53·8	54	97·4	54	140·8	54	184·1	54
54·5	55	98·1	55	141·5	55	184·8	55
55·3	56	98·8	56	142·2	56	185·5	56
56·0	57	99·5	57	143·0	57	186·2	57
56·7	58	100·3	58	143·7	58	187·0	58
57·4	59	101·0	59	144·4	59	187·7	59
+58·2	60	+101·7	60	+145·1	60	+188·4	60

In critical cases ascend.

INDEX